DEVELOPMENT/DIGRESSION DIARY ON INDIA

3D Companion Volume to Information India

1990-91

CONCEPTS IN COMMUNICATION
INFORMATICS & LIBRARIANSHIP
(CICIL)

For a list of titles in this series,
see pp. 142-44 at the end of the book.

ABOUT THE AUTHOR

S.P. AGRAWAL (*b.* 1929) was Member Secretary, Indian Association of Social Science Institutions and Director, National Social Science Documentation Centre, ICSSR. He was also the Vice-Chairman of the Asia and Pacific Information Network in Social Sciences (APINESS). In 1988 he was awarded the degree of Doctor of Philosophy in Education (Honoris Causa) by International University Foundation. Presently, he is Member, Round Table on Research in Reading, IFLA, and Member, International Committee for Social Science Information and Documentation. He is also Member, Board of Studies, National Archives of India, and Member, Hindi Advisory Committee, Ministry of Coal, Government of India. He has written extensively on various aspects of education, Indian polity, and library and information science.

Concepts in Communication
Informatics & Librarianship-45

General Editor: **S.P. Agrawal**

Development/Digression Diary of India

3D Companion Volume to Information India 1990-91

S.P. AGRAWAL
former Member Secretary
Indian Association of Social Science Institute, New Delhi

CONCEPT PUBLISHING COMPANY, NEW DELHI 110059

ISBN 81-7022-453-5

First Published 1993

Published and Printed by
Ashok Kumar Mittal
Concept Publishing Company
A/15-16, Commercial Block, Mohan Garden
NEW DELHI-110059 (India)

Lasertypeset by
Microtech Advance Printing Systems (Pvt.) Ltd.
H-13, Bali Nagar,
NEW DELHI-110015

GENERAL EDITOR'S INTRODUCTION

In this age of Information Explosion not unmixed with Mis-Information Explosion and Dis-Information Explosion every educated person is expected to be acquainted with basic facts about all aspects of one's national life. In view of the time factor, however, it is hardly possible to seek information on various subjects and topics of general interest from among the mass of material found in an ever increasing number of publications. And as such the best method to obtain the relevant information is to narrow the pavement by the help of reference guides which provide, in a capsule form, all mix of information relating to everyday life, the important events, issues and problems of one's country. The present Diary is intended to serve this purpose. Hopefully it will enable the users to find, quickly and expeditiously, and in a condensed form, information on topics of general interest as well as on certain specialized subjects.

Information India, which is being brought out annually since 1988, has been well received by its users. We have received suggestions from a number of our valued readers for incorporating information on certain topics which had been difficult to dig. This placed us under an obligation to include the information as required by the users. To fulfil this obligation, a lot of required material was collected from different sources. Later on, however, it transpired that inclusion of the material in question would make the *Information India*, which had already become tightly packed, unwieldy. In view of this, the material was re-packaged to bring out a separate volume in an alphabetical order.

Although this Diary is a companion volume to *Information India*, it is an independent volume in itself. It provides, in a concise form, information about a large number of topics and subject of general interest for non-specialist readers in an intelligible manner, notwithstanding that it also includes information on specialised and technical subjects. We are, however, conscious of the limitations and the drawbacks it suffers from, that are inevitable in view of the circumstances under which it has been prepared – while on some items information provided is scanty, some topics of general interest, which normally have to be included, have, however, not found a place in it. Depending upon the usefulness to be judged by encouraging user response, the Diary may continue as an annual publication, and if that be so, all-out efforts would be made to overcome the weaknesses and to fill the gaps.

Suggestion are welcome.

S.P. AGARWAL

CONTENTS

General Editor's Introduction

Diary

A

ACCIDENTS

Air

Since 1980, out of 18 accidents to Indian registered Boeing aircrafts in scheduled operations, only in five cases flight crews did not adhere to laid down proceedings, and action was taken against the flight crew found blame worthy.

Human error and technical failures were among the main factors responsible for the "significantly high number of major accidents" in the Indian Air Force which resulted in a loss of Rs. 419 crore over the past five years.

Human errors accounted for 35 to 42.85 per cent of major accidents between 1983-88 while technical failure or malfunctioning of components led to 25 to 50 per cent of these accidents.

Contrary to the general belief, accidents caused by bird hits ranged between 7.5 and 14.28 per cent only.

The percentage of "major accidents" and "minor accidents" involving fighter aircraft was 68.25 and 59.31 per cent respectively, while trainer aircraft and helicopters accounted for 13.27 and 12.32 per cent of major accidents.

The total loss of aircraft in respect of fighters ranged between 52.63 and 83.33 per cent while it was 7.14 and 17.14 per cent respectively in the case of helicopters and trainer aircraft.

Similarly fighters, trainers, and helicopters were involved in 62.96, 18.52 and 12.96 per cent of the fatal incidents.

The Comptroller and Auditor-General of India (CAG) traces the factors responsible for human error accidents and incidents to "lack of knowledge or skill and experience and attitudinal faults, besides inadequacy in training processes".

Monitoring of post-accident activities with reference to the recommendations made by the courts of enquiry on operational, technical or administrative aspects "was not very effective" in Air Headquarters till April 1988.

See also Indian Air Force: IAF Jet Crash.

Railways

See Derailment.

Road

At least 53,252 persons were killed in road accidents all over the country during 1990.

Uttar Pradesh recorded the highest number of accident deaths (6,693), followed by Tamil Nadu (6,663), Maharashtra (6,452), Andhra Pradesh (5,077), Karnataka (3,946), Gujarat (3,504), Rajasthan (3,465), M.P. (2,764), West Bengal (2,404) and Bihar (2,145).

In Delhi at least 868 persons were killed in road accidents during this year as compared to 808 persons last year.

While DTC, State and private buses accounted for 229 road deaths, 240 people were killed by goods vehicles and 88 by private cars. There were a total of 4,038 vehicles involved in accidents during the year.

World Bank figures on road accident statistics place India at the fourth position with 19.70 accidents per 1000 vehicles. Mauritius tops the list with 73.75 accidents per 100 vehicles, followed by Hong Kong at 48.15 and Kuwait at 36.19. The lowest figure is

---

Source: Parliamentary Proceedings as reported in *Hindustan Times* and *The Hawk*, if not otherwise mentioned.

for Sweden where 4.1156 accidents took place for 1000 vehicles.

ACQUIRED IMMUNO DEFICIENCY SYNDROME (AIDS)

See also HIV Infection.

National AIDS Control Programme was started in 1987. The programme had been supported by the external funding agencies such as WHO and USAID.

The activities undertaken so far under the National AIDS Control Programme include:

(i) Organising a surveillance system for determining the prevalence and infectivity trends in groups practising high risk behaviour;

(ii) Promoting safety of blood and blood products by establishment of testing facilities for screening blood in blood banks on a zonal basis;

(iii) Health education;

(iv) Formulation of guidelines for hospital infection control;

(v) Strengthening of 13 regional hospitals and training of para-medical staff in diagnosis and management of HIV infected patients; and

(vi) Implementating focused activity plans in States and undertaking where a relatively larger number of HIV infected persons have been detected on the basis of surveillance.

World Bank aid to fight AIDS

The World Bank has agreed in principle to provide assistance to control AIDS in the country through a project with an estimated outlay of $65 million.

While the base cost was estimated at $55 million the remaining amount was intended for cost escalation and other unforeseen contingencies.

AIDS Cases

As many as 44 full blown cases of AIDS and about 2,167 sero-positive cases were reported till end March 1990 in the country.

AIDS Deaths

At least 48 persons have died of full blown AIDS in the country in the last three years. While 15 deaths occurred in Maharashtra, eight each were reported from Delhi and Madras since 1989. Manipur accounted for four deaths while two each occurred in Kerala, Pondicherry and Goa.

Government has identified 10 medical colleges in the country where medical care would be provided to HIV infected persons and AIDS cases

AFFORESTATION

The target for afforestation which includes social forestry is 5,50,000 hectares are coverage of public lands, 225 crore seeding distribution under farm forestry during 1990-91. The compensatory afforestation target for Sardar Sarovar project is 40,700 hectares and compensatory afforestation has been done over 3,016 hectares till November this year.

AGMARK

See Standardisation.

AGRICULTURE

See also Farm Sector.

Exports

The exports of agricultural commodities have gone up by 28.71 per cent during 1990-91. These exports amounted to Rs. 5,740.71 crore during 1990-91 compared to Rs. 4,460.04 crore in 1989-90. The Gulf region is a major market for agricultural exports and this appreciable increase is against the background of the recent Gulf war. Tobacco exports in 1990-91 are better than the best exports recorded during 1981-82. These exports amounted to Rs. 263.69 crore in 1990-91 compared to Rs. 172.03 crore in 1989-90. Plantations (tea and coffee) exports went up from Rs. 1088.78 crore in 1989-90 to Rs. 1335.49 crore in 1990-91, an increase of 22.26 per cent (Tea figures are up to Feb'91). Marine products exports shot up from Rs. 634.76 crore in 1989-90 to Rs. 890.40 crore in 1990-91. It has exceeded the target of Rs. 850.00 crore which was itself considered ambitious. Cotton exports have contributed significantly to increase in agricultural exports this year. It went up from Rs. 80.27 crore in 1989-90 to Rs. 752.76 crore in 1990-91. Cashew exports have also gone up from Rs. 363.92 crore in 1989-90 to Rs. 434.81

crore in 1990-91. Rs. 529.02 crore worth of cereals were exported in 1990-91 compared to Rs. 434.15 crore in 1989-90. However, exports of spices went down from Rs. 274.36 crore in 1989-90 to Rs. 235.15 crore in 1990-91 because pepper prices were considerably lower in the international market as compared to previous year. Besides, there was a sizeable shortfall in pepper production also. Pepper accounts for roughly 2/3rd of total exports of spices. Exports of extractions have also gone down mainly because of failure of groundnut crop in Gujarat and decline in value realisation of exports.

(Source: *Yojana*, No. 11, June 30, 1991)

Labour

There were 64.4 million agricultural labourers in the country including 8.9 million marginal workers. The number of female agricultural labourers was 28.5 million which constituted 44.85 per cent of the total female workers.

AIDS

See Acquired Immuno Deficiency Syndrome.

AIR INDIA

During the period January to June, 1990, Air India earned an estimated profit of Rs. 28.50 crore, according to provisional figures while Indian Airlines suffered an estimated loss of Rs. 32.50 crore (provisional).

Performance of Air India

(Rs. in crores)

	1986-87	1987-88	1988-89
1. Operating revenue	890.32	995.87	1150.85
2. Operating expenditure	843.78	1001.90	1084.26
3. Total revenue	925.46	1023.95	1205.11
4. Total expenditure	895.30	1067.36	1161.80
5. Operating profit	+ 46.54	- 6.03	+ 66.59
6. Net profit/loss	+ 30.16	- 43.41	+ 43.31

The Comptroller and Auditor-General (CAG) of India has severely criticised frequent cancellation of flights by Air India, loss of income on account of endorsement of tickets to other airlines and charging low charter price for VVIP flights.

There was a net loss of Rs. 7.3 crore on account of cancellation of flights during 1987-88. On the same basis, during the summer season of 1988-89, a net loss of Rs. 6.86 crore was estimated.

Air India did not have a regular system of assessing the revenue loss due to cancelled flights, except in the case of VVIP flights. The loss of revenue to the airline on account of cancellation of flights is worked out on the basis of actual revenue earned on similar flights as reduced by the cash cost. Cash cost incurred on operation of flights includes aircraft fuel, landing fees and handling charges, material consumption, navigation charges, operating and cabin crew expenses, food and cabin services, etc.

During 1987-88, one A-310 aircraft was impounded by Nigerian authorities for 58 days from August 25, 1987 to October 21, 1987 for allegedly carrying contraband goods. It had seriously affected the normal schedule of flights of Air India. It resulted in cancellation of 78 single services. ATKM to the extent of 8.7 million was lost and the net loss of revenue due to the detention of aircraft at Lagos was estimated at Rs. 211.20 lakh.

For passengers transferred to other carriers, Air India pays them normal fare laid down in the IATA tariff irrespective of what has been earned by Air India on those tickets. In other words, the airline tends to lose heavily in respect of those passengers who were scheduled to travel by Air India on special fares or reduced fares and who were transferred to other carriers. The maintenance of systematic records at the Central office in respect of endorsements made by the stations had been done away with from March 1986 and it had been revived since July 1987. On the basis of available records, it was seen that from July 1987 to October 1988, 12,432 tickets had been endorsed to other carriers. Out of 12,432 tickets, 5,904 tickets, i.e., around 47 per cent were issued on special fares or reduced fares.

During October 1987, one aircraft from Middle East Airlines (MEA) was taken on lease for a total payment of Rs. 191.50 lakhs. It was used to operate 13 flights during the period from October 12, 1987 to October 23, 1987. This wet-leased operation resulted in a loss of Rs. 32.47 lakhs.

ALLOTMENT OF GOVERNMENT RESIDENTIAL ACCOMMODATION

The Comptroller and Auditor-General (CAG) of India flayed the Delhi Administration for delays in allotment of residential quarters to administration employees and for not taking timely action in eviction of people who did not qualify for retaining Government accommodation.

Against the demand of 2,064 quarters only 503 quarters were available for allotments as on March 31, 1989. Against the target of construction of 2,000 quarters at an outlay of Rs. 1,830 lakh during the Seventh Five Year Plan (1985-90), 416 quarters were constructed.

"While on the one hand, there was a long waiting list for residential accommodation, on the other, there were delays ranging up to 26 months in the allotment of 176 newly-built quarters reportedly due to lack of basic amenities like market, conveyance, school".

CAG criticised the department for delays ranging from two to 15 months in the re-allotment of 211 quarters, stated to be due to administrative reasons.

64 quarters were laying vacant ranging from the date of their construction, one from March 1983 and 44 from March 1987. The resultant amount of loss was Rs. 21.20 lakh by way of rental and avoidable payment of house rent allowance.

Serious lapses in control mechanisms to initiate timely action for eviction of persons who ceased to be eligible for Government accommodation resulted in number of quarter remaining under unauthorised occupation.

There was no effective watch on regular recovery of rent and as on March 31 last year an amount of Rs. 11.32 lakh was outstanding in 68 cases test checked. "These included Rs. 9.75 lakh recoverable from 58 Government servants who had already retired, died or were transferred".

Ground rent of Rs. 34.94 lakh due for the period from October 1986 to January 1990 had not been recovered from three oil companies.

Instances of considerable delays ranging up to 62 months in allotment to government servants of units vacated by earlier occupants and up to 17 months for newly constructed units have been highlighted by the Comptroller and Auditor General of India.

Certain gross anomalies were there in the allotment of dwelling units to Government employees. Due to acute dearth of residential flats, a Government employee could ordinarily get allotment of accommodation only before he rendered service ranging between 17 and 27 years.

Large number of *ad hoc* allotments were being made and 48 per cent of the total allotment in Type V in 1988 was on an "out-of-turn" basis. This further delayed the availability of accommodation for those entitled to regular allotment.

2,525 additional residential units became available for allotment during 1986 to 1988 as compared to 3,119 units during 1983 to 1985 and 12,814 units during 1981 to 1982.

Due to ineffective monitoring and control, cancellation of allotment in many cases of unauthorised occupation was unduly delayed. Delay in cancellation was more than 60 months in 26 cases.

See also Government Flats.

ALUMINIUM

See Mines and Mining Aluminium.

ASSAM OIL BLOCKADE

The Government suffered a loss of about Rs. 80 crore on account of the fortnight-long oil blockade by the All Assam Students Union (AASU).

Although the oil blockade was lifted and crude oil started flowing to the refineries, the ONGC *bandh* by the Subsagar District Students Union, an AASU affiliate, continued. The losses were estimated at Rs. 70 to 80 crore at current international prices.

Work at most of the production installations and drilling operations of ONGC in the State continued to be crippled by the ONGC *bandh*, which began on August 7.

AWARDS

Bharat Ratna

On the April 1990, the President, R. Venkataraman, conferred the country's highest civilian on Dr. B.R. Ambedkar, the principal architect of the Indian Constitution.

Booker Prize

The judges of the Booker Prize, Britain's richest award for fiction, has announced the six novels shortlisted for the 1990 award.

The books named for the awards are: "An Awfully Big Adventure" by Beryl Bainbridge, "Possession" by A.S. Byatt, "The Gate of Angels" by Penelope Fitzgerald, "Amongst Women" by John McGahern, "Lies of Silence" by Brian Moore and "Solomon Gursky was Here" by Mordechai Richler.

Coast Guard Medal

On the occasion of the Independence Day 1991, the President's Coast Guard medal for distinguished service has been awarded to Cmde (retd-in) J.P. Carneiro, Deputy Inspector General.

The President has also awarded two Coast Guard medals for meritorious service to Assistant Commandant B. Jacob and Navik Ajayvir Singh.

Dr Vikram Sarabhai Research Award

Dr. Vikram Sarabhai Research Award for 1989 by the Physical Research Laboratory (PRL) were awarded to the following 6 scientists for their contribution to different fields.

Professor L.M. Patnaik, Indian Institute of Science, Bangalore; Dr. Surendra Pal, ISRO, Bangalore; Dr. N. Kameshwar Rao, Indian Institute of Astrophysics, Bangalore; Dr Sarbeshwar Bajarbarua, Institute of Advanced Study in Science and Technology, Guwahati; Professor Sulochana Gadgil, Indian Institute of Science, Bangalore; and Shri M.C. Chandrasekhar, Scientific Secretary, ISRO, Bangalore.

The biennial award sponsored by Shri Hari Om Ashram of Nadia carries a medal and a cash reward of Rs. 8,000.

East-West Center Award

An Indian film, *Birth*, has won top honors at the recently concluded Hawaii International Film Festival.

The first directorial effort of cinematographer Shaji N. Karun, *Birth* won the *East-West Center Award* for the film that best promotes understanding among the peoples of Asia, the Pacific and the United States. Karun, who has been director of Photography for more than 30 films by some of India's top film-makers, also received the Eastman Kodak Cinematography Award in recognition of his lifetime achievement as a cinematographer.

Australian film critic David Stratton, who chaired the jury for the award, said of *Birth:* "Despite a deceptively simple story about a father anxiously awaiting a son who never returns home, it lingers in the memory long after it is over. The director seamlessly blends the personal story with an understated political message of oppression and the images are unforgettably beautiful".

(Source: *Span*)

Jamnalal Bajaj Awards

Dr. Ariyaratne of Sri Lanka would receive the international award for developing a model of socio-economic changes through mass participation.

Shri Tirath Ram of Himachal Pradesh will receive the award for his outstanding work on agricultural improvements, training for village industries and removal of untouchability. The award for application of science and technology in rural areas will go to Mr Shripal Davholkar of Maharashtra, while Mrs. Ratan Shastri of Rajasthan will be honoured for her untiring efforts for uplift and welfare of women and children.

(Source: *Statesman*, October 19, 1990)

Jawaharlal Nehru Award

President Robert Mugabe of Zimbabwe has been nominated winner of the 1989 Jawaharlal Nehru Award for

International Understanding.

The jury, headed by Vice-President Shankar Dayal Sharma, comprised Supreme Court Chief Justice Ranganath Mishra, Maharashtra High Court Chief Justice Chittatosh Mukherjee, Shri P.N. Haksar, Mr Nikhil Chakravarty, Prof N.G. Goray, Jawaharlal Nehru University Vice-Chancellor Dr M.S. Agwani and Indian Council for Cultural Relations Director General Veena Sikri.

Shri Mugabe had been in the forefront of Zimbabwe's freedom struggle from his early youth and was imprisoned for his activism. He studied and appeared for his law degree from London University while under detention.

Jnanpith Award

Eminent Kannada writer V.K. Gokak has been selected for the 26th Jnanpith Award for the year 1990 for his outstanding contribution to the enrichment of Indian literature.

The award carries an amount of Rs. 2 lakh.

Shri Gokak, born in 1909, has given a new dimension to modern Kannada poetry and literacy criticism during the past 50 years. His first collection of lyrics marked a revolutionary change in Kannada poetry. He rose to eminence after his work '*Dyava Prithivi*' (1957) won him the Sahitya Akademi award.

Livelihood Honorary Award

US economist Alice Tepper Marlin founder and President of the Council of Economic Priorities (CEP) in New York, has won the 1990 Livelihood Honorary Award.

The jury has honoured Tepper for the "courage and vision in showing the directions in which Western economy must develop if it is to contribute to global security and wellbeing of humanity", says the Press release.

Three recipients, who share the 1990 Right Livelihood Cash Award of 120,000 dollars, are Africa's peasant leader Bernard Ledea Ouedraogo, Israel's human rights lawyer Felicia Langer, who for over two decades defended Palestinians from gross injustice in the Israeli military courts, and Colombian agricultural workers' organisation, "Association de Trabajadores Campesinos del Carare".

Moortidevi Award

The Eighth Moortidevi Award for the year 1990 will be presented to Muni Shri Nagaraja (born 1917), renowned scholar in Hindi, Sanskrit, Pali and Prakrit and an exponent of ancient Indian culture for his monumental work *Agama aur Tripitaka: Ek Anushilan.*

Both thematically and from the contemplative point of view, this work of Muni Shri Nagaraja presents an original vision highly valuable from the historical and philosophical angle. Muni Shri Nagaraja has introduced a new style and mode of presentation by comparing both the Jaina and Buddha system.

National Citizens Award

Former Vice President B.D. Jatti, Minister of State for Agriculture Mullapally Ramachandran, film actress Ms Bhanu Priya, noted social worker Mrs Shakuntala Paranjape and eminent economist and Commerce Secretary Montek Singh Ahluwalia are among the National Citizens Award winners for 1990.

Other award winners include: Science and Technology Secretary, Dr P. Rama Rao, industrialist Dr Vinay Bharat Ram, Hakim Abdul Hamid (medicine), Mr S.L. Khosla, ONGC Chairman (public sector), Mr G.V.G. Krishnamurthy (jurisprudence), Mr V.N. Negi, IGP (law and order), Mr K.R. Ramamurthy (banking), Mr Mani Bhai Desai (rural development), Mr Gulshan Kumar (music), Mr Rajender Awasthi (literature), Mr Sidhartha Srivastava (electronic media), Mr B.S. Gupta (education), Mr Randhir Singh (sports), Mr P.K. Rami Reddy (entrepreneur), Mr Rishan Rapsang (youth development), Mr Manikan Ramaswamy (commerce), Mr Rishi (culture), Mr Turlapati Kutamba Rao (arts).

The awards are presented every year by the National Citizens Club, a voluntary organisation, to selected people for their distinguished service in their fields.

National Safety Award (British)

The Ghaziabad unit of the defence public sector Bharat Electronics Limited

(BEL) has been awarded the prestigious National Safety Award, 1989 of the British Safety Council.

The unit had an accident rate of 2.4 against the norm of five accidents laid down for the industry.

The BEL, Ghaziabad, has won this award for the second year in succession.

National Safety Award (Indian)

The Gujarat State Fertilisers Company (GSFC) polymer unit has bagged the Central Government's National Safety Award under the chemical and pharmaceutical sector for the year 1988.

This award has been given for the unit's excellent performance in terms of longest period of accident-free hours during the year.

Nobel Prize

The Nobel 1990 prize in medicine was jointly awarded to Dr Joseph E. Murray and Dr E. Donnal Thomas for their discoveries concerning "organ and cell transplantation in the treatment of human disease". The discoveries have enabled the development of organ and cell transplantation into a method for the treatment of human disease, the Karolinska Institute announced. Dr Murray discovered how rejection following organ transplantation in man could be mastered and Dr Thomas managed to diminish the severe reaction that the graft can cause in the recipient, the so-called "graft-versushost" reaction (GVH). In addition, Thomas could show that intravenously infused bone marrow cells were able to repopulate the bone marrow and produce new blood cells. Dr Murray successfully transplanted a kidney between homozygous twins for the first time. He pioneered transplantation of kidneys obtained from deceased persons and could show that patients with terminal renal insufficiency could be cured. The field was then open for transplantation of other organs such as liver, pancreas and heart. Dr. Thomas was successful in transplanting bone marrow cells from one individual to another. Bone marrow transplantation can cure severe inherited disorders such as thalassemia and disorders of the immune system as well as leukemia and aplastic anemia. American Jerome I. Friedman and Henry W. Kendall and Canadian Richard E. Taylor won the Nobel prize in physics on October 17, 1990 for making a break-through in man's understanding of the structure of matter.

During experiments at the Massachussets Institute of Technology (MIT) and the Stanford Linear Accelerator Centre in California, the three were able to find traces of quarks, the basic building blocks. Their experiments showed definitely that there were smaller nuclear particles than protons and neutrons. "The discovery was made when protons and neutrons were illuminated with beams from a giant electron microscope, a 3.2 km long accelerator at the Stanford Linear Accelerator Centre", where Taylor is affiliated. Dr. Friedman, 60, of Chicago, Illinois, and Dr Kendall, 63, of Boston, Massachusetts, are physics professors at MIT. Dr Taylor, 60, is a professor of physics at Stanford University in California. The winners were key researchers in a team which found clear signs that there exists an inner structure in the protons and neutrons of the atomic nucleus.

Shri Elias James Corey of the United States won the Nobel prize in Chemistry. Shri Corey, 62, a professor at Harvard University, was cited for "his development of the theory and methodology of organic synthesis".

Mexican author Octavio Paz, whose lyrical poetry explores the loneliness of man and draws on the rich imagery of Mexico's landscape, won the Nobel prize in literature. Paz also has written many distinguished essays critical of Mexican society, which have angered both the Right and the Left in his homeland. The Nobel Academy cited Paz, 76, for "impassioned writing with wide horizons, characterised by sensuous intelligence and humanistic integrity". Paz, who continued writing during a 25-year diplomatic career, was also praised for his "wide international perspective". "Paz's poetry and essays evolve from an intractable but fruitful union of culture, pre-Columbian Indian, the Spanish Conquistadore and Western modernism". One of the high points of his poetry was the long 1957 poem "Sun Stone", which was inspired by a huge Aztec calendar stone.

King Carl XVI Gustaf handed out gold Nobel prize medallions and diplomas at a ceremony for the 10 laureates in medicine, literature, physics, chemistry and economics on December 11, 1990.

A stand-in for Soviet President Mikhail Gorbachev accepted his Nobel Peace Prize in Oslo, Norway.

Olle Hansson Award, 1990

Jose Augusto de Barros from Brazil, Robert Lopez from Peru and Arturo Lomeli from Mexico have won the Olle Hansson Award for 1990, it was announced by the International Organisation of Consumers Unions (IOCU) (Regional Office for Asia and the Pacific). The Award recognises the work of outstanding Third World individuals in promoting the rational use of drugs.

Dr Barros coordinates the Latin American office of Health Action International (HAI), a network of citizens, health and consumer groups working for the safe, rational and economic use of drugs. His work includes an analysis of women as the principal target of drug manufacturers in their advertising because they are seen as primarily responsible for their family's consumption of drugs.

Dr Lomeli is a founding member of HAI and has promoted and investigated local and regional aspects relating to consumer health issues.

Dr Lopez's campaigns for rational drug use in Peru have contributed substantially to consumer awareness, particularly in the rural areas.

The Award is named in memory of Olle Hansson, a leading Swedish paediatric neurologist and renowned advocate of patients injured during the 1960s by the anti-diarrhoeal drug clioquinol. He is also known as a powerful campaigner against the reckless marketing practices of drug multinationals.

The Award, which carries a prize of US $1,000 each, is managed by an international panel coordinated by the IOCU, which is a federation of consumer organizations dedicated to protecting and promoting consumer interests worldwide through research, information and education activities.

(Source: *Development Forum* 18(5) Sep-Oct 1990)

Raja-Lakshmi Award

Dr Nataraja Ramakrishna, an exponent of temple dances of Andhra Pradesh, has been chosen for the Raja-Lakshmi Award for 1991. The Raja-Lakshmi literary award for 1991 goes to eminent Telugu poet, Dr Boyi Bhimanna.

Rajasthan Shree Award

Rajasthan Diwas Samaroh Samiti conferred the prestigious "Rajasthan Shree" award to Mr Vijay G. Kalantri, eminent industrialist and social worker. Mr Kalantri, a Prawasi Rajasthani settled in Bombay has contributed significantly to strengthening the bonds of Rajasthanis residing outside the State of Rajasthan. The award is in recognition of his services.

Shram Awards

Seventeen persons, including a woman and a visually handicapped artisan, have been selected for the Prime Minister's Shram Awards for 1991, given to workmen for outstanding contribution to production and for showing exemplary zeal in the discharge of their duties.

Of the three Shram Bhushan Awards, two have been shared by teams of four and five workers, respectively. One individual winner is Mr M.M. Singh Saini of the Steel Authority of India Limited, Rourkela Plant. The Shram Bhushan Award carries a cash prize of Rs. 50,000.

Three persons have been selected for the Shram Vir Awards, each carrying a cash prize of Rs. 30,000. While one women has been selected for the Shram Devi Award, three men have been selected for the Shram Shri Awards. These Awards carry a cash prize of Rs. 20,000 each.

Mr M.M. Singh Saini, the Shram Bhushan Awardee, has been selected for his important contribution towards the smooth running of the captive power plant at the Rourkela steel complex. Modification of slurry discharge pots and elimination of choking of the hopper ash discharge system of the electrostatic precipitator have been credited to Mr Saini.

Mr Saini's innovations effected a direct saving of Rs. 1.31 crore during 1990-91 on consumption of furnace oil. The annual power production went up from 78 MW to 92 MW during the period.

A team of five workmen from Nuclear Fuel Complex of the Electronics Corporation of India Limited (ECIL) has also been selected for the Shram Bhushan Award. They are Mr Bala Venkata Rama Mohan Rao, Mr Narendra Roy, Mr Mohammad Sarwar Miya, Mr T Sriramula and Mr Reddipally Koteshwar Rao.

This team of workmen performed the challenging task of indigenous fabrication of the cold reducing tube mill in the Nuclear Fuel Complex, leading to increased production rate as well as better quality of tubing. An official release says that the mill fabricated by this team has been working smoothly and giving a higher rate of production with better quality than those of the imported version.

Another team selected for the Shram Bhushan Award consists of four workmen from the Vindhyachal Super Thermal Power Project. They are Mr S.K. Dubey, Mr S.K. Sikdar, Mr R.D. Ravte and Mr H.S. Sindhu. The team has been selected for executing the job of repairing the damaged generator rotor collar, which was considered beyond repair even by the representatives of the Russian suppliers of the equipment.

Procurement of a new rotor from the USSR would not only have taken more than a year but also entailed a loss of Rs 81.14 crore per annum in terms of power generation. This team of workmen worked round the clock for 22 days and repaired the rotor.

The three Shram Vir Awards have gone to Mr Yad Ram of the 505 Army Base Workshop, Delhi Cantonment, Mr Gopikasanthanam Kamalavasagam, a master technician with the Hindustan Aeronautics Limited, Bangalore and Mr Kshetra Pal Singh, a fitter with Bharat Heavy Electricals Limited, Hardwar.

The Shram Devi Award has gone to Mrs Kiran Wilson of BHEL, Hardwar for coming up with a bending device for armature coils that solved a serious problem of delay in the manufacture of large size DC motors by reducing the rejection rate to zero. These motors are critically required for the extension of the Bhilai and Vizag steel plants.

The three recipients of the Shram Shri Awards are Shri H.B. Lingaiah, a visually handicapped artisan with BHEL's Electroprocelains Division at Bangalore, Mr Bijayananda Moharana, a technician with the National Aluminium Company Limited at Damanjodi in Orissa, and Mr Ghouse Mohiddin, a skilled fitter with the South Central Railway workshop at Hubli in Karnataka.

Special Citations

The National Chemical Laboratory (NCL) has won two special citations from the Indian Chemical Manufacturers Association (ICMA) for developing the technology for indigenous manufacture of acrylates, thereby saving precious foreign exchange for the country.

Teachers Award

From 1987 onwards, the award money had been increased from Rs. 2,500 to Rs. 5,000 and since 1988, the number of awards had also been increased from 186 to 300.

There is no proposal to further increase the number of awards for teachers and the value of award money.

UNESCO Human Rights Prize

The United Nations Educational Scientific and Cultural Organisation had awarded its 4,000 dollars annual Human Rights Prize to Czechoslovak President Vaclav Havel.

Mr Vaclav Havel's outstanding role as a human right militant and philosopher, the radiation of his work in this field and its major influence on the historical transformations in Central and East Europe made him a natural winner of the award.

AYODHYA TRAGEDY

Sixteen persons were killed in the incident at Ayodhya on October 30 and November 2, 1990.

B

BANGLADESH REFUGEES

237 tribals from Bangladesh entered Tripura between November 1989, and March 1990. During the same period, 181 persons had sought refuge in Meghalaya, 708 in Assam and 320 in Mizoram but they were either pushed back or returned on their own accord.

BANK AND BANKING

Frauds

See Frauds in Banks.

Profits

The published profits of foreign banks operating in India has gone up from Rs. 116.73 crore as on end-March 1989 to Rs. 184.08 crore as on end-March 1990. Published profits of the Indian nationalised banks has gone down from Rs 254.16 crore to Rs. 249.60 crore between these periods.

BATTERY BUSES

The Comptroller and Auditor General of India has pulled up the Delhi Energy Development Agency (DEDA) for not making viable an environment-friendly bus service which it launched in June 1985.

The battery bus scheme conceived as a measure to control pollution, had sustained losses totalling Rs. 124.98 lakh between 1985-89.

Instead of earning a profit of Rs 15 per bus per day the buses were running at a loss primarily due to failure of the DEDA to provide suitable back-up facilities, infrastructure and proper supervision.

The battery buses were intended to provide a noiseless mode of transport for the congested areas of the Capital, thereby providing a solution to the dual problems of growing traffic and noise and air pollution.

The agency, which had a fleet of 101 buses as on Dec 31, 1989, had projected an average run of 60 kms per day per bus. However, the average performance per day, per bus ranged from a dismal 8.43 km to 17.14 km.

60 to 70 per cent of the buses had been off the road since the inception of the scheme due to the non-availability of parking facilities and charging stations.

The poor performance of the buses was attributed by the agency to frequent failure of the chopper controller fitted in each but which is an imported electronic item.

Though the number of buses graduall increased, steps were not taken to provide adequate parking facilities and to solve the problem of failure of chopper controllers.

With a view to increasing the number of kilometres run by the battery buses from 50 to 100 per shift, it was decided to add an extra battery to enable quick replacement. Of the 24 buses modified by BHEL, extra batteries were provided to only three.

See also Delhi Energy Development Agency.

BEGGARS

There are an estimated 10,000 beggars in the Union Territory of Delhi.

A number of vocational trainings like cutting and tailoring, soap-making, bandage cutting, weaving , laundry, match-making, etc., are being imparted to the beggars for their rehabilitation.

The Bombay Prevention of Begging Act, 1959 as extended to the Union Territory of Delhi and Delhi Prevention of Begging Rules, 1960, are in force in Delhi.

BETEL LEAVES

The estimated average value of annual production of betel leaves during the last three years is Rs 700 crore. Betel leaf was not covered in the price support policy for agricultural commodities.

BHABHA ATOMIC RESEARCH CENTRE

Unprecedented downpour resulting in

a landslide temporarily damaged the basement of one laboratory at the Bhabha Atomic Research Centre, Bombay. The equipments at the Centre was affected following the downpour but was now in working condition. As a measure of caution, there was a proposal to provide a multiple barrier, strengthen the existing drainage system and relocate some equipment.

BHARAT HEAVY ELECTRICAL LIMITED (BHEL)

India's first HVDC line established full rated power flow of 100 MW over 196 km DC line between Barsoor in Madhya Pradesh and Lower Sileru in Andhra Pradesh on August 20, 1990.

Funded jointly by Departments of Heavy Industry, Electronics and Atomic Energy, the entire system design, development, engineering and manufacture has been done by BHEL engineers without a redo of any design feature.

The two sites — Barsoor in Bastar district of Madhya Pradesh and Lower Sileru in Andhra Pradesh — are located in remote and backward areas of the country.

This Research and Development Project has been implemented under the leadership of Bharat Heavy Electricals Limited assisted by the two concerned State Electricity Boards APSEB and MPEB forming the major core group of project team and helped by a number of other organisations such as Central Power and Research Institute, NTPC, CEA, BARC, Central Soil and Materials Research Station, Research and Academic Institutions such as Indian Institute of Science, IITs as well as a number of Government Departments like Departments of Heavy Industry, Electronics, Power, Atomic Energy, Science and Technology and Planning Commission.

For this project, the Bhopal Plant of BHEL has designed and developed for the first time in the country, converter transformers, smoothing reactors, capacitors for AC filter banks.

BHARAT NIRMAN PLAN

Bharat Nirman — an all India organisation to promote and take up constructive programmes in all walks of life — has submitted an action plan to the Union Government for the mobilisation of an additional Rs 50,000 crore. It has been suggested that the Government should announce another scheme for the declaration of the unaccounted money both by the resident as well as by the non-resident Indians (NRIs). There are several thousands NRIs who would like to bring their money through legalised channel provided the Government makes it possible to do so by imposing only one point entry tax at the rate of 50 per cent. The Government should not enquire about the origin of the same money and exempt it from all kinds of taxes. This alone would lead to the mobilisation of around Rs 30,000 crore.

Apart from this, the Government should sell out at least 24 per cent of the equity holding from the leading profit making public sector companies like Maruti, Air Lines and Steel Authority of India. There by without losing the controlling power in these concerns the Government could raise around Rs 10,000 crore. It has also been suggested to the Government to sell out the various public sector undertakings either as going concerns or their assets which do not fall under the strategic, defence or public utility items. This could raise an additional resource of Rs 10,000 crore, the memorandum suggested.

The financial institutions and the nationalised banks should be asked not to invest in the private sector industries until and unless they are in the priority or strategic sectors. This would lead to huge savings in resources which in turn could be utilised for the improvement of irrigation system, rural housing, warehousing, cold storages and in the other priority and agro industries sector.

BHOPAL GAS VICTIMS

The process of disbursement of interim relief of Rs 200 per month per person in the 36 severely affected municipal wards of Bhopal, as per the scheme of Central Government, is in progress. As on August 23, 1990, bank accounts in favour of 1.79 lakh beneficiaries were opened in this context.

BIRTH RATE

The birth rate in India declined from 33.7 in 1980 to 31.5 per 1000 population in 1988.

BLIND

The estimated number of both-eye-blind persons in the country is 11.92 million. According to an All India Survey conducted by the Dr R. P. Centre for Ophthalmic Sciences, AIIMS 1986-1989, the prevalence of both-eye-blind persons is 1.49 per cent.

BONDED LABOUR

State Governments have reported the identification of 2,55,608 bonded labourers as on March 31, 1990, of which 2,22,935 have been rehabilitated. Accordingly to available information, the total number of bonded labourers identified and freed countrywide as on May 31, 1990 was 2,44749.

BOOK INDUSTRY AND TRADE

Publishing Houses

According to one of the surveys by NCAER, there are approximately 11,266 publishing houses which also include individual author publishers.

Book Production in India

According to the National Library, Calcutta, during 1989-90, the titles produced in India were 17,040.

Publications Received by National Library, Calcutta, under the Delivery of Books Act

Year	Titles
1983-84	14,202
1984-85	14,591
1985-86	15,720
1986-87	16,790
1987-88	16,886
1988-89	16,635
1989-90	17,040

As per the *UNESCO Statistical Year Book 1989*, the position of Indian Publishing is eleventh.

Book Production in Select Countries

Country	Year	No. of Titles
1. USSR	1985	83,976
	1986	83,472
	1987	83,011
2. UK	1985	52,661
3. FRG	1985	54,442
	1986	63,724
	1987	65,670
4. Japan	1985	45,430
	1986	44,686
5. Rep. of Korea	1985	35,837
	1986	41,543
	1987	44,288
6. France	1985	37,860
	1986	38,701
	1987	43,505
7. China	1985	40,265
8. Spain	1985	34,684
	1986	38,405
	1987	38,302
9. Brazil	1985	17,648
10. Italy	1985	15,545
	1986	16,297
	1987	17,109
11. India	1985	11,660
	1986	12,543
	1987	14,965

Book Export

As estimated by the Book Panel of the Chemical and Allied Products Export Promotion Council (CAPEXIL), the export of books and publications had reached the Rs.300 million Mark during the year 1989-90.

Book Import

The conditions for import of books under OGL are:

1. Only books on educational, scientific and technical subjects are allowed. Fiction and Children's books if prescribed by any recognised institution as textbooks or recommended books are allowed.

2. No import of foreign editions of which authorised Indian editions are available.

3. Foreign reprints of Indian books are allowed with permission.

4. No import of unauthorised reprints of books published abroad.

Imports-Exports Gap

(in Million Rupees)

Year	Imports	Exports
1985-86	430	240
1986-87	499	260
1987-88	454.7	280
1988-89	1141	290

(Source: *Indian Book Industry*, May 1991)

BUDGET DEFICIT

See Fiscal Deficits; Government Budget Deficit.

BUSINESS HOUSES

See Industrial Houses.

C

CALCUTTA METROPOLITAN DEVELOPMENT AUTHORITY (CMDA)

The Comptroller and Auditor General's report on CMDA for 1987-88 reveals a surprising lack of administrative control on expenditure and use of raw materials and equipment. During 1977-80 the CMDA purchased seven spreader machines worth Rs 14.77 lakh for bituminous work in road construction. While two of the machines "were never utilised", the others were used for 27.64 days before 1982 and since then remained idle. Similarly three 2-ton capacity mobile garage cranes bought at Rs 2.5lakh in March 1983 were also left idle "without any utilisation".

The CMDA spent as establishment expenses an amount of Rs 18.34 lakh on divisional office in Howrah in connection with the modification, renovation and replacement of the gas distribution system. It, however, spent a mere Rs 2.41 lakh on the actual work. "Thus, the CMDA had maintained an office establishment for more than three years without having sufficient work load."

Between July 1983 and April 1984 the CMDA bought prestressed concrete pipes worth Rs 91.28 lakh for the Baranagar-Kamarhati after treatment plant and carried the pipes to the work site spending another Rs. 3.56 lakh. "In July 1986," "it was decided not to use such pipes because these were unable to withstand the pressure of heavy vehicular traffic. The work commenced in April 1987 with mild steel and cast iron pipes and the concrete pipes were kept unutilised on Barrackpore trunk road and Bangur stockyard."

There is a curious instance of the disappearance of 355 tons of bitumen worth Rs 12 lakh. Two firms were given a contract in May 1985 to carry 3,500 tons of packed bitumen from the suppliers' depots in Howrah and Haldia to the CMDA godown. The estimated value of the bitumen to be transported was Rs 1.09 core and the carriage cost was Rs. 2.09 lakh. Out of the total 1,113 tons lifted by the firms between Oct. 1985 and May 1986 from the suppliers' depots, 355 tons or 30 per cent never reached the CMDA godown. The failure to exercise necessary control over the movement of bitumen led to the cumulative short delivery of 355 tons bitumen worth Rs 12 lakh.

CANCER PATIENTS

There are about 1.5 to 2 million cancer patients in the country and almost half a million new cases occur every year.

CAPITAL GOODS

The import of capital goods during 1989-90 up to January 1990 was Rs 8,082.70 crore while during 1988-89 up to January 1989 it was Rs 5,525.32 crore.

CASHEW NUT

Exports

Export of cashew kernels from India between January and March 1990 registered a 14 per cent growth 1990.

During this period, exports totalled 8,110 tonnes, compared to 7,133 tonnes in the corresponding period the previous year. Export earnings also appreciated by 10 per cent despite the fall in prices, totalling Rs 65.44 crore, as against Rs 59.75 crore the previous year.

Intake by the Soviet Union, Czechoslovakia, Romania and the German Democratic Republic showed an appreciable increase and they together accounted for half the exports. Exports to Japan, Hong Kong, Kuwait and the Federal Republic of Germany also showed improvement.

Export prices, both for wholes and broken grades declined during the quarter. The unit export price for wholes declined by 4.2 per cent from Rs. 86.14 per kg. in January-March 1989 to Rs 82.51 per kg in January-March 1990. It declined from Rs 65.91 per kg to Rs

59.01 per kg for brokens, a fall of 10.6 per cent.

Exports in March 1990 totalled 2,858 tonnes, an increase of 32 per cent over the 2,160 tones exported in March 1989. The earnings realised were also higher at Rs 22.88 crore, an increase of 27 per cent over the Rs 17.99 crore earned in March the previous year.

The Netherlands accounted for 40 per cent of the exports, followed by the USSR with 28 per cent and Japan with nine per cent. Exports to USA were only 123 tonnes (four per cent).

During the month, prices for brokens moved down by 0.8 per cent from Rs 65.33 per kg in February to Rs 60.12 per kg while prices for wholes increased marginally by 0.13 per cent from Rs 81.75 per kg to 81.86 kg.

Exports in the financial year 1989-90 totalled 44,857 tonnes and was valued at Rs 360.24 crore 32 per cent higher in quantity and 30 per cent in value than exports in 1988-89. Direct exports to major markets — Canada, Czechoslovakia, GDR, Kuwait, Netherlands, New Zealands, Singapore and USSR—were much higher during the year, and lower in the case of Australia, FRG, Hong Kong Japan, Saudi Arabia, UAE, UK and USA.

The average exports price realised during the review period was only Rs 80.33 per kg 1.14 per cent less than the Rs 81.26 per kg realised the previous year.

A quantity of 553 tonnes of cashew kernel liquid, valued at Rs 0.43 crore was exported during March 1990 as against the 48 tonnes valued at Rs 0.05 crore, exported in March 1989.

Total exports from January to March 1990 amounted to 1013 tonnes, valued at Rs 0.80 crore, against 558 tonnes valued at Rs 0.49 crore in the same period in 1989 showing an increase of 82 per cent in quantity and 63 per cent in value.

Exports in the financial year 1989-90 totalled 3,014 tonnes, valued at Rs 2.53 crore, as compared to 3,034 tonnes valued at Rs 2.49 crore the previous year. The marginal rise in earnings was due to the higher unit export price realised during the year, which stood at Rs 8.379 per tonnes as against Rs 8,133 per tonne the previous year, a 2.4 per cent increase.

During March, 1990 of 6,716 tonnes of raw cashewnuts worth Rs 10.50 crore was imported into India, compared to 4,078 tonnes worth Rs 5.70 crore in March 1989. The total import from January to March 1990 were 24,302 tonnes valued at Rs 36.85 crore — an increase of 73 per cent in quantity and 79 per cent in value over the import of 14,046 tonnes, valued at Rs 20.64 crore from January to March 1989. Imports in 1989-90 were 52,475 tonnes valued at Rs 72.44 crore while in 1988-89, 37,078 tonnes worth Rs 51.89 crore were imported.

(Source: *Hindustan Times*)

Two lakh tonnes of raw cashew nuts was exported during 1990-91.

CATARACT

See Eye Diseases.

CEMENT

Cement Production

(in Million tonnes)

Period	Production
1970-71	1.43
1980-81	1.86
1985-86	3.29
1986-87	3.48
1987-88	3.74
1988-89	4.40
1989-90	4.50
1990-91	4.90

CENTRAL ADMINISTRATIVE TRIBUNAL (CAT)

The Government proposes to set up a new bench of the Central Administrative Tribunal (CAT) at Nagpur and also wants to increase the number of courts in a few benches already established.

As many as 37,648 cases were pending before the principal bench and four other benches of CAT.

CENTRAL GOVERNMENT HEALTH SERVICES (CGHS)

The Comptroller and Auditor General of India has found serious lapses in the working of CGHS which covers over 37 lakh people throughout the country.

The scheme, which started with 16 allopathic dispensaries covering 2.23 lakh people in Delhi in 1954, has been extended to 16 States with 221 allopathic units and 13 allopathic clinics and 71 units of the Indian system of medicines.

The annual expenditure of the scheme had increased from Rs 29.63 crore in 1984-85 to Rs 55.20 crore in 1988-89 and the average cost per card holder had increased from Rs 375 to Rs 650. While the expenditure had increased over the years, the amount spent on medicines was proportionately less. The average cost per card holder varied widely from station to station.

Despite the recommendations of the Estimates Committee, the Government did not compile and publish the dispensary-wise information on cost per card-holder. Audit analysis of the expenditure in respect of selected dispensaries showed that the cost per beneficiary in high-income areas was relatively more. In Delhi, the average annual expenditure on medicines per beneficiary was Rs. 9 in Srinivaspuri and Rs 65 in North Avenue.

The contribution from the beneficiaries covered just 13 per cent of the recurring expenditure. Despite the recommendations of he Fourth Central Pay Commission, the rates of contribution from Government servants prescribed in 1954 had not been revised.

The Seventh Five Year Plan targets of opening of new units were not likely to be achieved.

Several shortcomings were noticed in the working of dispensaries. The criteria for opening dispensaries were not observed. In Delhi, the number of families attached to a dispensary varied from 1,547 (President's Estate) to 17,432 (Kingsway Camp). The dispensaries were working a single six hours shift, though it had resulted in inconvenience to the beneficiaries. The deployment of medical and paramedical staff conformed neither to the sanctioned strength nor to the recommended norms, resulting in uneven workload. Fifty per cent of the doctors whose work was test-checked, attended to 50 patients against the norms of 75 patients a day.

The family folders system, which would give the total history of a patient at a glance and put a restraint on malpractices, had not been introduced though recommended by the Estimates Committee.

The expenditure on specialist services had increased from 9.5 to 11.3 per cent of the total expenditure between 1984-85 and 1988-89. Though the attendance in dispensaries had declined, the number of cases referred for specialist treatment had increased. There were no facilities for emergency beds, the requisite number of specialists were not provided, X-ray and ECG equipments were not functioning or remained closed for want of films, etc.

Dental clinics at Ahmedabad, Allahabad and Kanpur did not function for up to five years for want of dental surgeons.

In the procurement and supply of medicines, several irregularities in provisioning resulting in excessive stocking, short supply necessitating local purchases at extra cost, receipt and supply of substandard medicines and non-disposal of expired medicines were noticed. In Delhi alone, the additional expenditure on local purchase of medicines was about 33.1 lakh during 1987-88 and 1988-89.

There was no analytical system of quality control in the purchase of Ayurvedic and Unani medicines. The procurement of Unani medicines fell short of the requirements due to delay in the finalisation of contracts. Cases of non-availability of homoeopathic medicines for long periods were also noticed.

In Delhi, 75 out of 111 units did not have proper accommodation. In other cities also, 95 out of 194 units were housed in private buildings. Despite an adverse notice by the Estimates Committee in 1981-82, no targets were fixed for the construction of CGHS buildings during the Sixth and Seventh Plans.

The reporting system and monitoring by the Director, CGHS, showed that though several returns had been prescribed, the compilation of statistical data was in arrears for over two years, and the data published or compiled were inaccurate. The scheduled inspections of the dispensaries were not conducted and complaints were not monitored.

(Source: *Hindustan Times*)

The restructuring of the CGHS is still under consideration of Government. One of the components of the restructuring programme relates to changes in the system for dispensation of medicines.

CHILDREN

Child Labour

The number of child labour (0-14 age group) in the country was 13.64 million as per the 1981 census. The highest concentration of child labour is in Andhra Pradesh at 1.9 million, followed by Madhya Pradesh at 1.7 million and Maharashtra at 1.6 million. According to a survey conducted during July 1987-June 1988 child labour in the country was estimated at 7.6 million.

See also Girl Child Labour.

Desolate Children

The government has raised grants for maintenance of desolate children from Rs 150/ per month per child to Rs 250/ per month per child from April 1990.

Runaway Children

There has been an increase in the number of children who ran away from their homes in Delhi during the last three years.

While 1759 children including 426 girls ran away from homes in 1989-90, the number was 1666 children including 345 girls in 1988-89, 1064 children including 290 girls ran away in 1987-88.

Poverty, neglect, destitution, deprivation etc. are among the main reasons for children running away from their homes. A variety of welfare schemes were being implemented for providing care and protection to the neglected, disadvantaged and weaker sections of society, he said.

27 voluntary organisations are running Children's Homes in Delhi in which 2,991 destitute, orphaned and vagabond children are being maintained.

CLAMPING OF VEHICLES

Clamping of wrongly parked vehicles was stopped by Delhi Traffic Police following a court order.

COAL

During the year 1989-90 the coal production in the country was 200.87 million tonnes as against the target of 20950 million tonnes, the short-fall being mainly in ECL, BCCL and SCCL. The target of production has been fixed at 221 million tonnes (CIL 194 m.t., SCCL 22.50 m.t. and others 4.50). The plan allocation is Rs 2305 crores (CIL Rs 2031 crores SCCL Rs. 274 crores). During the first quarter of the current year, i.e, April-June, 1990 coal production from CIL was 93.2 per cent of the target. SCCL's achievement during the first quarter was 76.5% of the target.

During 1989-90 the Government sanctioned 26 coal projects including 15 Advance Action Schemes.

Coal production during 1990-91 was 211.73 million tonnes, which was 10.84 million tonnes higher than in the previous year.

The cess collected by coal companies by Coal India in 1989-90 was Rs 995.47 crore. This cess was being passed on to the concerned State Governments.

The gross sales of Coal India Ltd were Rs 6,248.43 crore in 1990-91 according to provisional figures. In 1988-89 and 1989-90 it was Rs 5,600.09 crore and Rs 6,278.32 crore, respectively.

The Working Group of Planning Commission on coal and lignite has estimated the total demand of coal in the terminal year of the Five Year Plan, i.e., 1994-95 at 320 million tonnes. To meet this demand, a production target of 304.45 million tonnes has been suggested for CIL and SCCL mines and 5.89 million tones for TISCO, IISCO and DVC mines.

COAL INDIA LIMITED (CIL)

The Comptroller and Auditor-General of India has criticised (CIL) for incurring and infructuous expenditure of Rs 1.34 crore on the ill-planned purchase of a second-hand 45-year-old plant without first assessing its technical suitability, financial viability and feasibility of location, and finally disposing it of at a loss without deriving any benefit from it.

After obtaining the formal sanction of the Government in February, 1981, the second-hand Jabalpur thermal power plant of JESU was purchased from the Madhya Pradesh Electricity Board (MPEB) in April 1981 at the negotiated price of Rs 74.85 lakh plus Rs. 8.98 lakh sales tax. The plant was earmarked for installation at Eastern Coalfields Limited. However, no project report was prepared before investment in the project.

The company, on the recommendations of the adviser to the Bureau of Public Enterprises, decided to engage the services of Disbergarh Power Supply Co. (DPSC), a subsidiary of Andrew Yule and Co. (AYC), for the installation and operation of the power plant. In April 1981, Andrew Yule & Co. presented a proposal for connecting the proposed power station with the existing distribution network of the DPSC. It was also proposed that there should be a close CIL-ECL-DPSC tie up with a possible equality participation. However, the proposal was not acceptable to the Government, and as such alternative agencies including BHEL, were contacted for the dismantling, transportation and installation of the JESU power plant.

After a strip examination of the plant in August-September, 1983, the AYC got a project report prepared which envisaged a total capital investment of Rs 982 lakh. It was proposed that ECL/CIL should grant an interest free loan of Rs 732 lakh to the DPSC in addition to their own contribution of Rs 250 lakh. The proposal was agreed by the SCL and the CIL. In May, 1985, Andrew Yule & Co, however, informed the company that the project for the installation of the 45-year-old thermal plant was not acceptable as it was found neither technically suitable nor financially viable, owing to the proposed withdrawal of investment allowance and increase in the price of cement, steel, freight etc. Thereafter, the possibilities of using the JESU plant or parts thereof by other coal companies were explored but those were not considered technically feasible.

The board of directors of the company in its 68th meeting held on July 15, 1985, decided to dispose of the plant. Accordingly, tenders for the sale of the plant were floated by Eastern Coalfields Limited in April 1986. Out of three tenders received, the highest offer was less than 50 per cent of the reserved price of Rs 104 lakh. Moreover, the cheque for earnest money received from the highest tenderer was dishonoured by the bank. The plant was, therefore, with drawn from sale and was retendred in January 1987. The highest bid of Rs 76 lakh quoted by a private party against this retender was accepted and the sale order issued. The amount of Rs 76 lakh was deposited by the party between November 1987 and July 1988, though as per the sale contract the payment should have been completed by Dec. 31, 1987.

In the meantime, Eastern Coalfields Limited had to pay the MPEB Rs 30.54 lakh as the upkeep and maintenance charges of the pland for the period from 1981-82 to November 1988. The Board directed that ground rent at Rs 1,000 a month and security charges of actuals from Oct. 14, 1987, to July 19, 1988, i.e. up to the date of full payment, be recovered from the buyer. The amount is yet to be recovered from the company.

COFFEE

The crop of 1989-90 at 117,800 tonnes available for marketing in 1990-91 is far below the 1988-89 crop of 215,000 tonnes which was available for marketing in 1989-90. The target set for coffee export for 1990-91 is Rs 400 crore.

COINS

Import

During 1987-90, 1230 million pieces of stainless steel coins in the denomination of 50 paise, 25 paise, and 10 paise were imposted at a total cost of Rs 17.09 crore. The production of coins in the Government mints was 2945 million pieces in 1987-88, 3183 million pieces in 1988-89 and 2844 million pieces in 1989-90.

COIR

Exports

During 1988-89, 24979 tonnes coir worth Rs. 33.32 crore was exported as against 25147 tonnes valued at Rs. 32.20 crore in 1987-88. During 1989-90 coir exports touched a new high of Rs. 40.18 crore.

This has been possible because Government has taken various steps to boost coir exports. These include sending trade delegations, conducting market studies and market research, participation in fairs in major markets, quality improvement and grant of export incentives.

Coir is a major cottage industry in India providing employment to about 5 lakh people. Coir production is mainly concentrated along the coastal regions of Kerala, Tamil Nadu, Karnataka and Andhra Pradesh are also emerging as important coir producing States. This industry caters to the export market in coir yarn, coir floor coverings like mats, mattings and carpets, and ropes etc. For the development of coir industry, Government had set up the Coir Board which is a Statutory Body under the Coir Industry Act, 1953.

(Source: *Yojna*)

COMMITTEES/COMMISSIONS/ PANELS/WORKING GROUPS

See also Haskar Panel Report; Industrial Relations: Arbitration and Legislation; Industrial Relations: Workmen's Compensation and Benefits; Pollution: Saikia Panel; Ports: Free Port.

COMMUNAL RIOTS

A total of 2,025 people were killed in 62 "major communal riots" in the country during 1989, 1990 and upto July 1991. The communal violence in various parts of the country in 1989 and 1990 to the heightening of tension because of the Babri Masjid-Ram Janmabhoomi controversy.

The year-wise break-up of the riots is: 1989 (1174 people killed in 18 riots), 1990 (693 people killed in 28 riots) and 1991 (up to now) (158 people killed in 16 riots).

In 1989, the maximum number of six riots took place in Bihar in which a total of 1,061 people were killed. Of these, 984 were killed in one single riot in Bhagalpur. In the same year Bihar was followed by Rajasthan, where five major communal riots took place and 30 people lost their lives.

In 1989 in Uttar Pradesh only two riots took place — at Badaun and Varanasi — claiming 30 and seven lives respectively.

The number of persons killed in major communal riots in the country increased from 33 in 1987 to 607 in 1989. Out of the 607 killed in communal riots in the country last year 491 were from Bihar.

COMPANIES

Non-Government, Non-Financial Public Limitd Companies

The performance of selected 622 non-Government, non-financial public limited companies each with a paid up capital of Rs one crore and above indicated a substantial revival of the business activities in 1988-89 reflecting a sizeable overall improvement in sharp contrast to the slackening trends witnessed during the previous two years.

According to a recent study made by the Reserve Bank of India, there was overall improvement in the key indicators particularly in respect of value of production, sales, profits, savings and capital formation. The growth rates in value of production, sales and value added an gross profits recorded substantial increases in 1988-89 as compared to with the growth rate in the previous years. The year 1988-89 witnessed a sharp rise in net savings as against a decline in the previous year. Important profitability ratios also recorded marginal increases during the year and the number of profit making companies increased to 452 in 1988-89 from 431 in 1987-88. The all-round good performance resulted in the average dividend rate registering a sharp rise in 1988-89. The number of companies declaring higher rates of divided of 20 per cent and above also increased to 214 from 173 in the previous year. However, the growth rates in gross fixed assets and net fixed assets were fractionally lower in 1988-89 compared to those in 1987-88.

The value of production and sales recorded substantially higher growth rates of 19.5 per cent and 18 per cent respectively during 1988-89 compared to 10.4 per cent and 10 per cent in 1987-88. The net value added registered a steep rise of 22 per cent in 1988-89 as against 8.5 per cent in 1987-88. Manufacturing expenses were marked up by 19.8 per cent in 1988-89 compared to 9.6 per cent in 1987-88 but its share in value of production only fractionally increased to 64.9 per cent in 1988-89 from 64.7 per cent in the previous year.

Remuneration to employees moved up by 13.7 per cent compared with 11.8 per cent in the previous year though its share in value of production declined fractionally to 11.3 per cent in 1988-89. Depreciation provision recorded a lower rise of 8.8 per cent in 1989-89 as against a steep rise of 21.5 per cent recorded in the previous year. Its share in the value of production also declined to 4.4 per cent during the year from 4.8 per cent in the previous year.

Gross profits recorded an impressive growth rate of 33.5 per cent in 1988-89 when compared to the mere 3.3 per cent rise in 1987-88. Interest payments increased at a higher rate of 17.2 per cent compared to 13.2 per cent in the previous year. The growth rates in operating profits, per-tax and post-tax profits registered steep increases of 60.5 per cent, 55.8 per cent and 74.5 per cent, respectively in 1988-89 in sharp contrast to the declines of 9.7 per cent, 4.5 per cent and 12.7 per cent witnessed in 1987-88. Retained profits increased by as much as 166 per cent as against a decline of 42.5 per cent in the previous year. Thus the selected companies indicated an appreciable all round improvement in profits in the year 1988-89 in contrast to the declines witnessed in the previous two years.

The higher growth rates of profit, in turn, were reflected in the profitability ratios. The return on total net assets increased to 9.3 per cent in 1988-89 from 8.1 per cent in 1987-88. The profit margin on sales also was higher at 9.9 per cent in the year under study compared with 8.8 per cent in the previous year. The return on equity recorded a sharp rise to 6.4 per cent in 1988-89 from 6 per cent in 1987-88. The divided rate was higher at 18.1 per cent in the year as against 15.5 in the previous year. Even after paying higher dividends, the share per cent of profits ploughed back into business in the post-tax profits sharply increased to 52.6 per cent from 34.5 per cent in the preceding year.

(Source: *Hindustan Times*)

New Companies

As many as 16,007 new companies limited by shares with an authorised capital of Rs. 7290 crore were registered under the Companies Act during the first nine months of 1989-90 (April-December 1989) as against 16,157 companies during the same period in the previous year. While the number of new companies registered during this period showed a marginal decline, authorised capital was significantly higher by 83.0 per cent over the same period.

A total of 65 companies limited by guarantee associations not for profit and 5 companies with unlimited liability were also registered during the period April-December 1989.

Maharashtra accounted for 22.2 per cent of the new company registrations during this period and 6.4 per cent of their aggregate authorised capital. Delhi accounted for another 18.9 per cent and five other States — West Bengal, Tamil Nadu, Andhra Pradesh, Gujarat and Uttar Pradesh accounted for over five per cent each of the company registration during this period.

These seven State together accounted for 78.1 per cent of the 16,007 companies during April December 1989 and their share in the aggregate capital was 92.4 per cent. Delhi accounted for the largest share 74.2 per cent of the aggregate authorised capital of the companies registered during this period. This was due to the fact that a Government company (the National Power Transmission Corporation Limited) having an authorised capital of Rs 5000 crore was set up during the period.

A total of 199 non-Government companies limited by shares are reported to have ceased to function during the year

1988-89 either by going into liquidation or being struck off under the Companies Act. The aggregate paid-up capital of these was Rs 15.53 crore. Another 246 companies limited by shares and having an aggregate paid up capital of Rs 27 crore were reported to have gone into liquidation during the first nine months of 1989-90 (April-December 1989), 157 companies with an aggregated paid up capital of Rs 9.8 crore were reported to have ceased functioning during the corresponding period of the previous year.

Of all the companies which ceased functioning during April-December 1989, 245 were non-Government companies and one was a Government company. As many as 39 of them were public limited companies and 207 were private limited companies having a paid up capital of Rs 7.7 crore and Rs 19.3 crore respectively.

An amount of Rs 3,828 crore was provisionally reported to have been raised during the year 1988-89 by was of paid-up capital by companies limited by shares. The paid-up capital of these companies during the first nine months of 1989 is a provisionally estimated at Rs 2332 crore.

Besides these Indian companies 421 foreign companies were also at work as on March 31, 1989. A total of 36 foreign companies established their place of business in India during April December 1989 and one comapany closed down its branch during this period. The total number of foreign companies as on December, 31, 1989 is 455.

During the year 1989-90, 178 non-Government non-financial public limited companies delivered to the Registrar of Companies copies of prospectus for raising an amount of Rs 1663 crore from the public. Of this 65.2 per cent of the total amount issued was offered to public, while the balance was earmarked for firm allotment of promoters.

The geographical distribution of companies at work shows that Delhi accounted for 17.7 per cent of the companies at work as on Dec. 31, 1989 and 43.1 per cent of the aggregate paid up capital. Maharashtra accounted for another 23.2 per cent of the companies at work. Three other States—West Bengal (14.4 per cent), Tamil Nadu(7.6 per cent) and Gujarat (6.2 per cent) had each over five per cent share in the total number of companies at work on this date. These five States together accounted for 69.1 per cent of the 1,92,999 companies and their share in the aggregate paid up capital 67.9 per cent.

Activity-wise distribution of these companies shows that 51.4 per cent of them were engaged in "processing and manufacture". Commerce (trade and finance) was the next largest major activity group accounting for 24.1 per cent and balance of 24.5 per cent of the companies were engaged in various other activities like agriculture, mining and quarrying, construction and utilities, transport, communication and storage, community and business services and postal and other services.

In addition to these companies limited by shares were also 324 companies with unlimited liability at work in the country as on Dec. 31, 1989. All these companies were non-Government private companies. While 253 of these companies were engaged in investment activity and 51 in trading, the remaining 20 were engaged in other activities, according to the annual report of the department of Company Affairs.

(Source: *Hindustan Times*)

COMPENSATION TO DELHI RIOT VICTIMS

An amount of Rs 15,25,97,000 has been paid so far as compensation to the victims of the 1984 Delhi riots.

See also Riot Widows.

COMPUTER HARDWARE AND SOFTWARE EXPORT AND IMPORT

Export and Import of Computer Hardware and Software

Period	Export Hardware	Export Software	Import of hardware and software clearance by Electronic Department for actual uses
1987	40.7	70.0	82.0
1988	137.8	101.0	238.0
1989	233.0	157.0	90.0

CONSTITUTION (66th) AMENDMENT BILL

On June 1, 1990, Parliament gave it approval to the Constitution (66th) Amendment Bill, 1990, seeking to bring all land reforms laws in the Ninth Schedule beyond litigation. The then Minister of State for Agriculture, described it as revolutionary step, which will provide relief to the lakhs of landless.

CONTAINER CORPORATION OF INDIA (CONCOR)

Container handling facilities for exports from Moradabad fulfil a long cherished need and desire of the entrepreneurs and exporters in that city and area around it.

The new container centre set up by the Container Corporation of India (CONCOR) at Pakwara, short of Moradabad on the Delhi-Lucknow National Highway No. 24, was commissioned on Sept 15, 1990 the honour of inaugurating it being ingeniously bestowed by CONCOR on one of the reputed master craftsmen (in 1980). Chotey Khan, who oversaw the first two containers out of the centre — one on way to Rotterdam (Netherlands) and other to Leeds (UK).

Moradabad, well-known throughout the country for exquisite craftsmanship of metal, particularly brass, has steadily emerged on the international trade map, its art metalware exports alone fetching for the country foreign exchange of the order of Rs 150 crore annually. Of a total of Rs 600 crores worth of handicrafts exports from all over the country during 1989-90 metal artware exports constituted per cent. In addition to brass artware, Moradabad exports tobacco paste (*hukka* tobacco) to the Middle East and some general goods, too. Most of the metal crafts from Moradabad go to the developed industrialised countries in North America, West Europe and Japan.

Handicrafts in the country constitute more than 2.5 per cent of the country's total foreign exchange earnings through export trade. This sector employs over three million artisans all over the country.

In the absence of container handling and customs clearance facilities, export consignments from Moradabad, metal artware as well as tobacco-paste — all containerisable cargo — has hitherto been moving out of Moradabad and adjoining areas by roads vehicles in break-bulk from their factories and manufacturing units, to be stuffed into international standard containers for export either at the ICD at Delhi or gateway ports of Bombay and Nhava Sheva. The old method led to delays in transit, frequent damage to cargo on account of multiple handling and exposure to inclement weather, besides causing considerable inconvenience and anxiety to the shippers.

The new arrangement will enable containers to be stuffed almost at the doorstep of the exporters, thereby achieving better loadability in the containers and thus saving in costs, particularly ocean freight, savings in packaging, better handling of cargo under the supervision of the exporters themselves, safe and speedy transit, in addition to conservation of energy and environment. The establishment of a container centre at Moradabad, with facilities for customs appraisal, survey and insurance facilities, customs house agents, documentation and other related aspects will further enable the exporters to derive full benefit of the now universally appreciated intermodal containersied cargo logistics.

To provide efficient, convenient and low cost facility to the shippers in the region, a road chassis system of operation is being institutionalised by CONCOR for the first time in the country. Empty container, both 20 ft, and 40 ft. in length, will be taken from Delhi on road trailers and moved to the container centre at Pakwara in Moradabad, equipped with covered storage space and ancillary facilities. Export cargo, stacked and prior-cleared by customs at the contained centre, will be stuffed in the empty container on chassis. The sealed container will thereafter be moved by road, covering about 160 km, directly to Tughlakabad rail container siding for transfer on to rail flat cars for onward despatch to gateway ports by express block liner trains. Consistent with CONCOR's objective of optimal inter-model development, this operation combines the strengths of both rail and road, conducive to overall fast, safe and economical service.

In order that the exporter does not have to approach different agencies at different places, CONCOR has provided for one-window service, quoting a composite rate for operations at the container station in Moradabad. For operations to remain cost effective to the maximum extent, most competitive rates are obtained by means of comprehensive open tender bidding and simplification of operational procedures and practices.

(Source: *Hindustan Times*)

CORE SECTOR

The infrastructural sector as a whole fared badly during 1990-91 and the uncertain prospect of its improvement in the coming months would affect the pace of development especially in industry and international trade, according to the Federation of Indian Chambers of Commerce and Industry (FICCI).

Electricity generation in 1990-91, FICCI grew only by 7.8 per cent during 1990-91 as compared to 10.9 per cent in 1989-90. The growth was also lower compared to 9.2 per cent in 1987-88 and the average growth rate of 8.3 per cent during 1980-81 to 1990-91.

The generation of thermal power (including nuclear) which accounted over 73.5 per cent of total generation increased only by 5.2 per cent as against the average growth rate of 11 per cent achieved over the last 10 years. The plant load factor of thermal stations has plummeted down to 53.5 per cent in 1990-91 as compared to 56.5 per cent recorded in 1989-90. The overall power shortage which has been steadily declining from 10.9 per cent in 1987-88 to 7.3 per cent in 1989-90 is estimated to have reached 11 per cent in 1990-91.

As regard oil and natural gas indigenous production of crude oil declined by 3.1 per cent during 1990-91 as against 12 per cent average rate increased achieved over the last 10 years. Also, crude throughput in refineries went up only by 2.8 per cent as against an average increase of 5 per cent posted over the past 10 years.

The coal production in 1990-91, was only 5.4 per cent as against 6.3 per cent average growth achieved over the last 10 years creating shortages of coal to the industrial units especially in the Western region. The coal availability to industrial sector has sharply declined over the last six months. Even the meagre supply of about 20 to 30 per cent of the requirement are also being disrupted frequently for one reason or the other.

Inadequate coal production is accompanied by bottleneck in the movement of coal by the Railways, coal movement to industrial units has sharply declined over the last six months and particularly during the past three months mainly because of the over-riding priority given to coal movement to the power house as also limited availability of rolling stock.

The performance of the shipping and ports are far from satisfactory. Cargo handling at major ports has registered only a marginal increase in recent years. The increase in cargo handling at major ports during 1989-90 was only 0.4 per cent as against the average rate of increase of 6.7 per cent registered over the past nine years. Apart from this, there are berthing delays for ship, longer time for loading and unloading which resulted in higher freight rates.

The shipping tonnage of Indian vessels registered only a negligible increase in 1990-91 from 5.76 million GRT in April 1987 to 6.03 million GRT in March, 1991. Of the total 6.03 million GRT, 0.52 million GRT is accounted for by coastal shipping capacity. There has been substantial scrapping of overaged vessels to the tune of 2.08 million GRT during the Seventh Plan which has led to curtailment of withdrawal of services in quite a few routes.

Road and road transport and telecommunications have also not made any significant improvement during 1990-91. The hike in petroleum products prices together with the stricter enforcement of carrying capacities with truck has led to steep increase in freight rates in the range of 20 to 40 per cent.

In the telecommunication sector, equipment shortages especially in switching and other critical components and accessories reserved for manufacture in the public sector and emerged as a major constraint.

(Source: *Hindustan Times*)

CORRUPTION

A total of 1,389 complaints of corruption were received against Income-tax officials by the Government and Central Board of Direct Taxes during the financial year 1989-90.

175 corruption cases were registered against group 'A' officers in the Government departments and public sector undertakings in 1990. The figures for 1988 and 1989 were 184 and 177 respectively.

CBI Probes

The CBI had taken up 119 cases for investigation against public servants during January to March 1990. As many as 37 cases were instituted against employees of the nationalised banks and 33 cases against the personnel of the railways.

See also Raids.

COTTON

Production Estimates

According to the present assessment, production is expected to be around 106 lakh bales as against the target of 115 lakh bales. The production during 1989-90 was 114.1 lakh bales.

COUNCIL OF SCIENTIFIC AND INDUSTRIAL RESEARCH CSIR

The Comptroller and Auditor General of India has indicted CSIR and two other science organisations for acquiring computers and leaving them idle.

A CSIR scheme to modernise its financial and administrative information system through computers blocked over Rs 82 lakh, while the computers remained unused.

There were complaints about components, hardware, software, training and non supply of manuals and in all Rs 82.87 lakh remained blocked.

Computers also remained idle at the Central Marine Fisheries Research Institute and the National Centre for Software Technology.

The CSIR had in 1987 ordered 40 personal computers and a larger system for its various laboratories and headquarters at a cost of about Rs. 100 lakh.

But there was no consensus among the laboratories either about the need or the capability of the computer system to be purchased.

Eight institutes felt that the system was not required, some refused to accept the system or refused payment after the delivery.

Even when the individual laboratories and a comparable system available to them, the new system was ordered for, which was not prudent and the objective of modernisation of financial and administrative information systems has not been achieved.

The Cochin-based Central Marine Fisheries Research Institute (CMFRI) lacked of planning in procurement of a computer system which led to blockage of Rs. 18 lakh for nearly two years.

The CMFRI ordered a computer in March 1986 at a cost of Rs. 18 lakh, but the site was not ready until March 1988 and it was installed with software only in May 1988 after a delay of 21 months.

The Department of Electronics (DOE) had prematurely released Rs. 144 lakh to the National Centre for Software Technology for importing a computer and constructing an advanced training facility.

The computer was procured in April 1987, but the land for the advanced training facilities had not been purchased.

Due to premature release of grants, the National Centre had unutilised funds of Rs. 74.71 lakh as on March 31, 1988 and the advanced training facility has not come up.

(Source: Press Trust of India *Release*)

CRECHES

About 280,000 working women are availing of creche facilities in 12,230 centres in the country. Rs. 13.72 crore spent on the provision of such facilities in 1989-90, Rs. 12.03 crore in 1988-89 and Rs. 11.95 crore in 1987-88.

Of the total plan allocation for the creche programme, 15 per cent was earmarked for the Scheduled Castes and 7.5 per cent for the Scheduled Tribes during 1990-91.

CRIME IN DELHI

41,245 crimes were committed in Delhi in 1990 as compared with 38,249 in 1989 and 34,443 in 1988.

See also Women Criminalism in Delhi.

CRUDE OIL

About 33 million tonnes of crude oil was produced in the country during 1990-91.

The entire demand of 3,884 million tonnes of petrol in the country was being met through indigenous production.

The country was also able to meet about 86 per cent of the demand for 56,781 million tonnes of petroleum products.

OPEC Crude Oil Production (January-June 1990)

(in '000 Barrels daily)

Saudi Arabnic	5402
Iraq	2992
Libya	1207
Gabon	270
Iran	2983
Indonesea	1175
UAE	2014
Algeria	750
Venezuela	2008
Dubai	410
Kuwait	1800
Qatar	380
Nigeria	1717
Neutral Zone	340
Abudhabi	1599
Equador	280

OPEC Crude Prices Rise

($ per barrel)

1986	13.10
1987	17.60
1988	11.50
1190 June	14.00
1990 July 27	18.00
1990 July 28	21.00
1990 August (Mid) 28.00	
Due to Gulf Crisis)	

Imports

The Government has approved the import of 19.6 MMT of crude oil and 9.62 MMT of petroleum products during 1991-92.

During 1989-90 the import of crude oil and petroleum products including from Middle East countries was to the tune of 26.03 million tones valued at Rs 6,318 crore.

The current use of natural gas was about 33 million cubic metres a day, which was being indigenously produced.

India, imported 8.7 million tonnes of petroleum products valued at Rs 4,660.21 crore in 1990-91 both under term contracts and through spot purchases.

Expenditures on Crude Oil Imports (Including Oil Products

(Rs. in crores)

1986-87	2773
1987-88	4015
1988-89	4378
1989-90	6440
1900-91 (estimated)	7440

D

DEBT

See Government Borrowings.

DEFENCE

Defence Expenditures in India

Year	Defence Exp.'000 crores'	GDP at factor cost '000 crores	Defence Exp. as percentage of GDP
1	2	3	4
1977-78	2,813	87,118	3.23
1978-79	3,060	93,724	3.26
1979-80	3,550	102,595	3.46
1980-81	4,091	122,427	3.34
1981-82	4,651.80	142,916	3.25
1982-83	5,408.30	158,851	3.40
1983-84	6,309.17	185,991	3.39
1984-85	6,660.51	208,577	3.19
1985-86	7,987.49	233,476	3.42
1986-87	10,477.45	259,055	4.04
1987-88	11,967.49	294,266	4.06
1988-89	13,341.02	351,724	3.79
1989-90(RE)	14,500	395,143	3.78
1990-91 (BE)	15,750.00	461,512 *	3.78
(RE)	15,750.00		
1991-92 (BE)	16,350.00		
			3.54

* Estimated

Source:

1. Defence Expenditure Data from Government of India, Defence Services Estimates for relevant years.
2. GDP data from *Economic Survey* (Government of India) for 1989-90 and 1990-91.

Defence Exports

Against a target of Rs. 69.71 crore involving direct and deemed exports, the total value of exports of arms and other allied weapons in 1989-90 was Rs. 80.27 crore.

Defence Personnel

81 defence services personnel were killed in encounters during the period August 1,1990 to July 31, 1991 in different areas of the country.

Government has provided necessary help to the families of those killed and as payments of liberalised pensionary awards to the dependents are continuing, it is not feasible to compute the exact amount spent on this account.

DELHI DEVELOPMENT AUTHORITY (DDA)

Delhi Development Authority had not yet been able to provide flats to all the people registered under the new pattern scheme, 1979, but concrete plants had been formulated to clear the backlog by 1994-95.

So far 106,669 of the 171,272 applicants had been allotted flats. The remaining applicants had not been given flats because of the non-availability for land and other infrastructural facilities.

40,116 allottees registered under the Rohini scheme were yet to receive their plots.

So far 41,176 residential plots had been allotted, with DDA developing 3,834 plots in 1989-91.

In 1986, 162 flats were allotted on an out-of-turn basis, 250 in 1987, 139 in 1988, 341 in 1989, 166 in 1990, 30 upto July 1991.

The functioning of the DDA had been reviewed periodically by several committees/agencies and the thrust of their recommendations was that the DDA should be divested of the extraneous functions conferred on it.

DELHI ENERGY DEVELOPMENT AGENCY (DEDA)

The report of the Comptroller and Auditor General of India strongly citicised Delhi Energy Development Agency. The facts reveal laxity in the functioning of the Agency marked by underutilisation of funds, non-recovery of dues and absence of any record of feedback on implementation of various projects undertaken and indicate that the Agency has miserably failed in achieving its aim.

The Agency was set up by the Administration in 1984. Its objective was to execute integrated rural energy programme and non-conventional urban energy programmes, various pilot projects of non-conventional and conventional energy sources and organising and promoting supplies of energy saving devices like solar cookers, wind mills, biogas plants. In addition a regular battery bus service was introduced by the Agency in 1985 as a measure to control pollution and provide noiseless mode of transport in the congested areas of the Capital.

The proposals for claiming grants from the Central Government and the Administration were not realistic. There were savings every year ranging from Rs. 8 lakh to Rs. 141 lakh. By March end last year Rs. 299 lakh was lying in fixed deposit.

Delhi Administration aimed to cover 70 per cent of the city's population that is about seven lakh families to adopt improved and non-conventional sources of energy at the end of the Seventh Plan. However there is no record to indicate the number of families which were adopting the improved and non-conventional sources of energy.

There has been no feedback on to the use of 631 solar cookers and 134 domestic solar water heating systems worth Rs. 13.35 lakh distributed by the Agency free of cost for demonstration purposes.

According to surveys conducted by the Agency during 1988-89 indicated that only 25 per cent of the 159 biogas plants, 31 per cent of the 115 domestic solar heating systems and 62 per cent of the 95 wind mills were working satisfactorily. The subsidy of Rs. 4.47 lakh paid on 119 biogas plants which were not working was unproductive. A sum of Rs. 15.71 lakh due from the beneficiaries of biogas plants and domestic solar water heating systems for the period 1984-89 was not recovered.

Out of five biogas engines purchased at Rs. 2.50 lakh between 1985-87 only one has been installed for demonstration and the remaining four have been lying idle.

No training was provided for the construction of biogas plants and there was no record on a training of demonstration of solar cookers although the government released Rs. 0.90 lakh and Rs. 3.50 lakh for the purpose respectively. Out of the 30 tropicultors along with carts purchased at a cost of Rs. 3.38 lakh in 1983-84 only two were sold and 22 tropicultors were defective.

Of the three windmill generators purchased at Rs. 4.80 lakh in 1985-87.

The Agency had also entered into an agreement with a doctor/inventor to set up a 300 kV power gas generation plant based on a technology the doctor had claimed to have developed for producing gas power from cellulosic wastes. The plant never came up and the expenditure of Rs. 10.87 lakh on the project was infructuous.

(Source: *Hindustan Times*)

See also Battery Buses.

DELHI RIOT VICTIMS

See Compensation to Delhi Riot Victims.

DELHI TRANSPORT CORPORATION (DTC)

In 1988-89 the loss was Rs. 9899.32 lakh, in 1989-90 it was Rs. 11,984.56 lakh, and in 1990-91 the provisional figure of losses was Rs. 19,748.35 lakh.

DELHI URBAN ART COMMISSION (DUAC)

The Delhi Urban Art Commission has stressed the need and importance of a comprehensive integrated development instead of a piecemeal approach.

In its 13th Report (1988-89), the DUAC says piece-meal planning lacks cohesion. In a few cases, the Commission advised the development of the schemes on principles of sound comprehensive and integrated concepts not to be pre-empted by arbitrary divisions of the overall pockets of land or by allotment of individual plots.

The Commission has also been insisting upon submission of properly worked out landscape schemes alongwith the proposals requiring its consideration.

In the case of Delhi City Museum, the whole area comprising the city wall, the Kashmere Gate, Nicholson Park, the proposed new park and the Museum complex should be comprehensively land-scaped in a befitting discipline.

While suggesting its development as part of a comprehensive scheme for the open area between Chandni Chowk and Railway Station, including the Town Hall gardens rather than being development in an isolated manner, the Commission advised that a comprehensive landscaping scheme for the open space between the railway station and the Chandni Chowk should precede the scheme for the proposed parking.

The Commission has identified the sign-board as an important element of street picture. While considering a few proposals of commercial complexes, the Commission emphasised that the graphics and sign-board system should be integrated into the scheme to avoid chaotic proliferation and haphazard display of sign-boards.

The report has emphasised the need for traffic study while considering proposals such as DTC depots and terminals and various large residential, commercial and industrial complexes.

In industrial complexes and wholesale markets the activities are very likely to give rise to a large quantity of refuse or waste, which if not cleared in time, can render the functioning of the complex miserable. The report notes that it is an important factor requiring prompt and adequate attention at the planning and concept development stage.

The Commission had the opportunity to examine a general pool housing development conceptual proposal at Mata Sundri Road, up to 12 storeys achieving an FAR (floor area ratio) of over 200 with a density of approximately 60 dwelling units per acre, thus utilising the relaxation in height and FAR provided in the revised comprehensive guidelines issued by the Government, but the parking provisions were not in accordance with those guidelines, the report observed.

High-rise structures adjacent to school-lane, flyover on south-ease side and other major roads on north-east side was likely to be affected by the air pollution caused by traffic. The high-rise block proposed at the junction towards Hotel Ranjit was also not considered desirable from the point of view of urban design consideration.

The organised open spaces were also inadequate for the large population of the complex. The liveability aspects in high-rise development, it was felt, also required a review. Adequate green buffer needed to be ensured between the development and the abutting major arteries, for minimising the effect of noise and air pollution. Traffic studies had also not been seriously thought of, the report says.

This demonstrated that the available relaxation in development norms cannot and should not be followed blindly. The relationship of the proposed complex to the existing surrounding development as well as its own impact should not be ignored.

(Source: *Hindustan Times*)

DELHI WATER SUPPLY AND SEWAGE DISPOSAL UNDERTAKING (DWSDU)

The Comptroller and Auditor General of India has criticised the Delhi Water Supply and Sewage Disposal Undertaking for slow pace of work on water treatment plants, extra expenditure and irregularity in awarding contracts.

The undertaking took up a Public Utility Scheme for augmentation of water treatment plants at Wazirabad in July 1986 for completion in two years, but the pace of construction had been slow as progress of different components ranged from zero to 75 per cent.

Tenders for supply of prestressed concrete and hume pipes were opened in August 1986 but supply orders were issued in June 1987.

Negotiations were held and the lowest tenderers got orders for a part of the supply. Other parties were given orders at higher negotiated. The undertaking incurred in avoidable expenditure of Rs. 45.52 lakh.

There was evidence to suggest that a tenderer had received the corrigendum well before the date of opening of tenders. "This undue favour to the contractor resulted an extra payment of Rs. 2.60 lakh."

Construction of a 100 million gallons per day water treatment plant, pumping sets, reservoirs at north Shahdara was taken up in 1980 for completion by March 1983.

"The plant had been commissioned in phases from June 1984. An expenditure of Rs. 23.47 crore had been incurred on the project up to 1985-86 and the work had not been completed."

The undertaking failed to properly

plan execution of allied works which led to payment of escalation of labour charges amounting to Rs. 7.99 lakh.

"Advances aggregating Rs. 232.10 lakh paid for deposit works to various agencies during March 1981 to October 1985 had been lying outstanding for four to seven years.

DEPARTMENT OF NON-CONVENTIONAL ENERGY SOURCES (DNES)

The Comptroller and Auditor General of India has pulled up the Department of Non-Conventional Energy Sources (DNES) for infructuous expenditure and non-achievement of goals in several projects. Four DNES projects failed to achieve their goals.

A project to set up a garbage incineration plant in Delhi to dispose off city waste and produce electricity led to an infructuous expenditure of Rs. 240 million.

The imported plant was intended to burn 300 tonnes of garbage a day and the heat generated was to be used to produce to two billion units of electricity per year.

DNES obtained a foreign loan of around Rs. 200 million, repayable in foreign exchange, for the project on the condition that it would be executed only through a foreign firm without doing any detailed reliable analysis of the garbage.

The garbage was later found unfit for the plant and it has been decided to wind up the project, rendering the whole expenditure infructuous.

DNES project to create indigenous capability for producing power using coal as fuel has still to realise its objectives, 13 years after its inception and after incurring an expenditure of Rs. 260 million.

The 'magneto hydro dynamics' (MHD) project selected coal as the fuel because of its abundance in India and because the technology once mastered for coal could be developed for natural gas but the reverse would not naturally follow.

Contrary to the mandate of commissioning a five mw MHD pilot power plant, the steering committee decided, in November 1984, that commissioning a plant to generate plasma could be taken as achievement of the objectives.

The pilot plant is being run on LPG instead of coal, contravening the approved project proposal.

The project which started in 1977 was to be completed by 1982. However, the objective of setting up a coal-based MHD plant for creating indigenous capability for economic generation of power through MHD technology is "still a distant goal".

In a wind pump demonstration programme in Andhra Pradesh, against a sanction of 150 wind mills, 73 were not procured and of the 77 procured, only 29 were reportedly working.

Rs. 7.49 lakh spent on the remaining 48 wind mills was unproductive.

Because of the large scale failure of the wind mill's in the demonstration programme, the beneficiaries were not coming forward for the wind mills.

Obviously user response and awareness in newer areas of energy sources would have adversely affected.

In another programme, the DNES committed itself to a project of using rice husk to run a 1.5 MW power plant without a proper appraisal of the project viability.

The DNES sanctioned the proposal of Punjab Agro Industries Corporation for installing the plant, initially expected to cost around Rs. 30 million, and released Rs. 10 million without proper appraisal.

But the corporation later decided to increase capacity to 2.5 MW and DNES was asked to increase its share to Rr. 50 million. DNES did not agree to this proposal and decided to withdraw financial support to the project.

As a result Rs. 10 million has been lying unutilised with the Government of Punjab for nearly three years.

(Source: *Press Trust of India Release*)

DERAILMENTS

There were 902 derailments during 1989-90 and 1990-91. The accidents took place mainly due to failure of Railway staff, equipment failure and sabotage.

DEVALUATION VIS-A-VIS EXPORT

To take maximum mileage for Indian

export, the policy of trading devaluation of rupee *vis-a-vis* currencies of India's major partners should be based on differential rates of inflation prevailing in India and abroad.

At present, the devaluation has been ad hoc in nature and does not always take into account the inflation rates prevailing in India and abroad. This has denied India the maximum advantage in export realisation resulting from devaluation.

If a policy of devaluation based on differential rates of inflation is followed, an exporter would be in a position to estimate the future value of rupee *vis-a-vis* other currencies. The expected rates of inflation in major countries are published regularly. This would help the Indian exporters to plan their strategies in advance to take advantage of the devaluation.

This suggestion is contained in a paper prepared by FICCI on Export Plan and Strategies for 1990-91. India has not promoted exports to countries where currencies have appreciated (or depreciated less) against rupee and import sourcing is done from those countries where currencies have appreciated in relation to rupee.

As for instance, India should try to import more from US because the rupee has depreciated least against the dollar and export more to Japan, West Germany and UK since rupee has depreciated strongly against these countries. Instead share of imports from the US has declined as have shares of exports to Japan and the UK.

The paper also argues for reduction in the import duties on capital goods since the devaluation of rupee has increased the Effective Rates of Protection (ERP) afforded to domestic manufacturers of machinery. Taking into account the value of machinery imported from various countries, the increase in the degree of protection to the domestic producers has been about 77 per cent in the period 1985-89.

To discern the thrust areas for export in the coming years, FICCI has evolved indices of 'relative penetration' of different primary export items in the world market. The index of penetration level of all merchandise trade has declined marginally from 0.46 per cent in 1982 to 0.45 per cent in 1987. A greater decline is discernible in the case of primary commodities. In the areas where India is losing out of the market share, specific policy package to boost the exports should be evolved.

On the specific export policy thrust required for individual items. In the case of marine exports, it might be necessary to remove the ban on trawling in monsoon months and also move towards fish farming. In the case of consumer electronics, FICCI maintains that given proper incentives, the export potential could increase from the present level of Rs. 50 crore to Rs. 500 crore.

(Source: *Hindustan Times*)

DIESEL

Approximately 21,000 TMT of diesel were consumed in the country during 1990.

DIESEL TROLLEY

Faced with the acute problems of unemployment and inadequate transport facilities, some people of Patti, a block of Amritsar District, have developed their own mode of conveyance called 'GHARUKA' — a diesel trolley. It is fitted with an eight-horse power engine, and consumes very little diesel compared to similar vehicles. It can carry about 50 passengers. Agriculture being the main occupation of the people, the trolley has proved very useful for quick transport of their agricultural produce to city market. It can also be used to operate tubewells and threshers. Each village of the Patti Block has more than 10 such vehicles. The owner of a trolley earns more than Rs. 200 per day. The 'Gharuka' is exempt from road tax and passenger tax payment. Besides, no registration and driving licence is required for their vehicle. The trolley costs about Rs. 30,000. Local people feel that it is a good source of employment with little investment. They think if loans are given to the unemployed youth from banks and other development agencies, the problem of unemployment can be solved to some extent.

(Source: *Yojana*, May 31, 1991)

DIRECTORATE OF ADVERTISING AND VISUAL PUBLICITY (DAVP)

The Comptroller and Auditor General has noted that DAVP incurred infructuous expenditure of Rs. 5.88 lakh on body building work of three vans meant for use by it in West Bengal and North East.

There were delays of three to five years in the completion of body building of the two chassis meant for Itanagar and Aizawal and the chassis meant for Calcutta was completed only in October 1989.

The infructuous expenditure has been mainly due to the delay in awarding the work to firms.

The directorate purchased three chassis for Rs. 4.75 lakh in November 1981 and March 1983 for use as exhibition vans for its mobile exhibition units at Calcutta, Itanagar and Aizawal. The Calcutta van was delivered in Delhi while the other two were given in Bhubaneshwar.

The fabrication work has awarded to a firm in Cuttack for Rs. 4.88 lakh in January 1985 on the basis of open tenders invited by the accounts officer of the DAVP regional office in Guwahati, to be completed within 60 days of receipt of chassis by the firm (February 1985).

"No valid reasons were forthcoming for the long delay in awarding the work."

While the two chassis meant for Itanagar and Aizawal were fabricated and delivered by the firm in March 1986 and March 1988, the third meant for Calcutta was withdrawn from the firm in March 1988 since it was not in a position to complete the work.

The chassis was brought to Delhi in March 1988 and the work awarded to a firm here in September that year at a cost of Rs. 2.45 lakh to be completed in 90 days. It was actually completed in October 1989.

According to the agreement with the Cuttack firm, DAVP was entitled to recover an amount of upto 20 per cent of the contract as liquidated damages for the delay. DAVP had paid Rs. 2.28 lakh to the Cuttack firm for body building on two vans and an amount of Rs. 1.43 lakh including security deposit of Rs. 0.45 lakh had been withheld.

The withdrawal of work of the chassis meant for Calcutta and is subsequent award to a Delhi firm resulted in extra expenditure of 0.98 lakh, which was recoverable from the Cuttack firm. This had not been recovered and DAVP had spent Rs. 2.72 lakh on the salaries of staff for the Calcutta van till October 1989.

The two completed vans reached Itanagar and Aizawal in March and May 1988, respectively, but commenced programmes only from March and September 1989.

The DAVP in July 1989 told the CAG that the units had remained idle since no field exhibition officer was posted, and later on, there were some problems of a technical nature due to which the vans could not be used.

An expenditure of Rs. 3.16 lakh was incurred on the staff before the vans were used for the programmes.

(Source: *UNI Release*)

DOORDARSHAN

The Comptroller and Auditor General of India has levelled serious charges against the functioning of Doordarshan. The electronic media did not have information about the total number of proposals received, contracts entered into and programmes completed each year.

This lapse should be viewed against the fact that Doordarshan spent Rs. 55.92 crore during 1985-90 on programmes produced by outsiders.

73 programmes contracted with the producers between February 1986 and March 1990, involving advance payments of Rs. 203.97 lakh, were pending for completion till October end 1990.

Fifty programmes received generally between June 1987 and March 1990 for Rs. 121.39 lakh were awaiting telecast up to October 1990, some of them due to changes in the political scenario.

Though producers were stated to be decided by the Director General on the basis of their experience or eminence, there was no panel of producers. The criteria for the selection of producers should be laid down clearly for the benefit of producers themselves, it said.

Though a particular producer failed to produce the programme within the stipulated period, producer was awarded a contract for further programmes. There was no systematic monitoring at Doordarshan kendras and no effective action was taken against defaulters or their surities.

The security deposit obtained from the producers was not deposited in Government account. Income tax amounting to Rs. 71.93 lakh had not been deducted at source from the payments made to the producers.

Doordarshan did not have a proper procedure for the sharing of copyright of programmes produced by outsiders.

Lack of coordination between two branches of Doordarshan resulted in the payment of an additional amount of Rs. 3.90 lakh on a programme earlier contracted for with the producer at a lower rate.

One programme already produced in-house and telecast thrice was awarded further to an outside agency, resulting in avoidable expenditure of Rs. 1.25 lakh.

A telefilm approved earlier as a sponsored programme for exhibition on payment of royalty at Rs. 0.32 lakh per telecast was purchased for Rs. 4 lakh. The telefilm had been telecast only twice.

In contravention of the prescribed schedules of payment, advance payment of Rs. 10.5 lakh was made in February 1989 for a six-part documentary serial for a total budget of Rs. 70 lakh. As the contract with the producer did not prescribe the scheduled date of completion, the serial had not been completed.

There was unjustified enhancement of Rs. 3.5 lakh to the original Rs. 5 lakh for the production of a film. The production of the film was also delayed by 14 months and the film was telecast after 31 months of its receipt, although Doordarshan was aware that "the print quality was not worthy of telecast".

The cost of a serial of seven episodes accepted by Doordarshan in November 1988 was enhanced from Rs. 12.6 lakh to Rs. 14 lakh. The serial was yet to be completed, though advance payment of Rs. 5.60 lakh was made in December 1988 and December 1990.

A telefilm contracted in February 1987 for which Doordarshan made advance payment of Rs. 2 lakh had not been produced so far.

There was a case of special treatment shown to a producer, saying Doordarshan agreed to enhance the cost of a documentary from Rs. 6.3 lakh to Rs. 8 lakh and gave 30 per cent foreign sale rights in favour of the producer, after the contract was signed. The producer was also allowed a more favourable schedule of payment.

An increase of 10 minutes in the total duration of a serial (10 episodes) led Doordarshan to bear an additional expenditure of Rs. 3.5 lakh when the contracted rate was only Rs. 1.75 lakh per episode of 30 minutes duration. Though all the episodes were stated to have been delivered in January 1990, their telecast was yet to be decided.

(Source: *PTI Release*)

Advertisement Revenue Earned

Doordarshan earned a gross revenue of Rs. 253.85 crore during 1990-91 from commercial advertisements. During 1989-90, the Revenue was Rs. 210.13 crore while in 1988-89 it was Rs. 161.26 crore. Doordarshan had earned Rs. 371.39 crore from advertisements and sponsorship programmes during 1988-89 and 1989-90.

The television serial *Mahabharat* was the single largest advertisement earner, as the amount of revenue earned during this serial was a staggering Rs. 57.42 crore.

This was followed by *Ramayan* which netted Rs. 24.40 crore in advertisement revenue. While the serial *Buniyad* fetched Rs. 8.87 crore. *Uttar Ramayan* earned Rs. 7.33 crore. The serial *Mirza Ghalib* earned Rs. 1.03 crore and *Tamas* Rs. 0.13 crore.

News

Doordarshan has allotted around 713 hours of transmission time per month for news while agricultural programmes get about 101 hours.

Up to 10 per cent of total transmission time could be made available for commercial advertisements, though only 2.5 per cent of the time is at present being used for the purpose.

Films

164 feature films were telecast by Doordarshan Kendra, Delhi, on the national network during last one year i.e. from August 1989 to July 1990.

During the last three years, Doordarshan telecast 542 feature films in various Indian languages and dialects in the national network and on Delhi and its linked transmitters.

Of these, 56 feature films in different dialects and six of these were in Bhojpuri.

Advertisement Telecast

The overall time devoted to the telecast of commercial advertisements had not progressively increased during the last three years and continues to be around 2.5 per cent as against the ceiling of 10 per cent prescribed for the purpose.

Coverage

The radio and TV services are at present available to 85 per cent and 57.7 per cent by area and to 95.7 per cent and 78.7 per cent by population respectively. With the completion of the ongoing schemes, the radio and TV coverage shall be increased.

DOWRY DEATHS

See under Women

DRINKING WATER

See also Technology Mission on Drinking Water

Indian Problem Villages

As many as 9201 villages in the country were facing problems concerning drinking water, with the highest concentration in Orissa (1615), followed by Meghalaya (1383) and Himachal Pradesh (1107). 5218 villages with no internal water sources remain to be provided with safe drinking water facilities in the country. Of the 161,772 villages identified on April 1, 1984 as in problem villages, 156,504 had been given safe drinking water fully or partially by August 1, 1991.

World View

People in rich countries take drinking water for granted. People in poor countries — 77 per cent of the global population struggle to get this basic necessity of life and even then have to worry whether it is safe to drink.

What is being done to get safe water to them? What should be done? Can all people hope for clean, drinking water available at their doorsteps? These are some of the issues to be examined and debated by experts, policy makers, activists and NGOs in New Delhi, 10th through 14th September, 1990 at the Global Consultation of Safe Water and Sanitation for the 1990s, which is to assess the progress during the 1981-90 "Water Decade" and to debate what should follow.

The New Delhi meeting is sponsored by United Nations Development Programme and hosted by Government of India. Over 100 countries, numerous International Agencies and non-Government Organisations are to be represented in the New Delhi meeting.

The New Delhi meeting is the culmination of meetings at the country, regional and inter-regional lovol during the Decade. The outcome of the Delhi Consultation will be brought to the attention of the UN General Assembly at its 45th Session scheduled in the Fall of 1990 when it will consider the Report of the Secretary General on the assessment of the International Drinking Water Supply and Sanitation Decade (IDWSSD) and the strategies for the 1990s.

The Agenda for the global consultations along the following main themes:

1. Bridging the Gap: Water supply and Sanitation in National Planning.
2. Creating a Safe Environment for Better Health: Water Resources, Sanitation and the Environment.
3. Extension, Communications and Community Management: Strategies towards the Social Goals.
4. Building National Capacities: People and Institutions.
5. Preparing for the Future: Development, Application and Transfer of Technologies for Water Supply and Sanitation.
6. Financing the Water Supply and

Sanitation Sector.

7. Building the Collaborative Framework: Coordination at the Global and Regional levels.

(Source: *Hindustan Times*)

DRUGS AND PHARMACEUTICALS

Substandard Drugs

Twelve drugs have been declared substandard by the Government analyst, Drugs Control Laboratory, Maharashtra State under section 18(A) (I) of the Drugs and Cosmetics Act, 1940.

The name of the sample with batch numbers are: chloramphenicol palmitate oral suspension, 164 — oxytetracylline injection IP 3590 25 per cent dextrose UNJ IP 4938 L 2 — sodium chloride and dextrose INJ IP (DNS 0.2 and 5), 3194, PDLYTE-M 2905-A, water for INJ IP, MIF Lot 481, Max Cilin syrum ampicillin for oral suspension IP, D-170 — sodium bicarbonate INJ NFI SB 1 mfg date April 89 — Chemox capsules, 119056 — Nelsulf dusting powder, 8905 — five per cent dextrose INJ IP D-6090 — five per cent dextrose INJ IP, D-17389.

TABLE D.1: **Seizures of Drugs : State-wise Data, 1989**

(Quantity in kg)

Name of State	Total No. of cases	Opium	Heroin	Ganja	Hashish
Andhra Pradesh	677	43.000	2.254	3110.564	—
Assam	12	74.939	1.562	—	—
Andaman & Nicobar	9	—	—	8.750	0.018
Arunachal Pradesh	740	1.750	—	8756.783	—
Bihar	146	0.800	0.065	15318.290	621.367
Chandigarh	29	21.633	32.777	8.750	3.885
Delhi	464	211.917	199.955	70.090	3308.976
Goa	30	0.066	0.001	0.116	2.091
Gujarat	75	97.887	16.573	26.995	29.283
Haryana	658	1125.585	2.784	57.216	33.655
Himachal Pradesh	41	9.180	—	—	65.485
Jammu & Kashmir	47	1.020	76.705	—	243.520
Karnataka	230	2.210	13.252	221.612	—
Kerala	207	25.101	23.348	2248.148	1171.277
Madhya Pradesh	196	1038.740	5.714	147.100	—
Maharashtra	478	128.307	1103.374	1306.375	396.067
Manipur	144	2.646	15.062	8297.775	—
Mizoram	44	—	3.137	8.890	—
Meghalaya	7	—	5.858	262.000	—
Nagaland	15	—	1.816	352.200	—
Orissa	3	0.600	1.000	720.000	0.400
Pondicherry	3	—	0.031	0.026	—
Punjab	166	777.664	758.917	27.340	25.483
Rajasthan	211	886.825	369.248	430.572	1311.085
Tamil Nadu	1276	76.380	30.603	11409.695	1.003
Uttar Pradesh	1218	200.610	33.897	1202.142	901.615
West Bengal	88	58.340	16.200	471.595	5.500

TABLE D.2: **Agency-wise Data, 1989**

(Quantity in kg)

Agency	No. of cases	Opium	Hero-in	Ganja	Hashi-sh	Metha-qua-lone	Mor-phine	Coca ine	Pheno-bar-bitone
N.C.B.	109	45	291	7842	494	326	10	—	720
D.R.I.	12	—	12	1156	84	193	—	—	—
Customs	570	184	515	200093	2553	165	1080	—	—
Cen. Excise C.B.N.	65	615	7	2	5	—	4	—	—
B.S.F. & Other Central Agencies	60	8	946	304	505	—	—	—	—
State Police	5434	1553	504	9241	6931	152	—	—	—
State Excise	86	63	102	509	6	—	—	—	—

Drug Addicts

About 2.27 lakh drug addicts were registered with various de-addiction, counselling and after-care centres during 1990-91.

The number of women and female students who are drug addicts is extremely small.

Union Welfare Ministry spent Rs. 36.54 lakhs on Drug campaign during 1989-90.

No census of drug addicts has been undertaken and hence the number of drug addicts in the country could not be indicated. However, Government was taking various measures to control the situation.

The steps taken include improvement in the inter-State and inter-agency enforcement coordination, training of customs, police and other enforcement officers and improvement in international, regional and bilateral cooperation in drug enforcement by ratifying the UN convention of 1988 and signing bilateral agreements with several countries like USA, Mauritius and Afghanistan.

See also Sports: Indian Sportsmen and Nadrolone.

Drug Policy Review

Government's decision to include an additional list of 21 bulk drugs under the purview of price control gave rise to expectations of a review of the drug policy of 1986 and the Drug Price Control Order (DPCO) of 1987. Both the leading organisation of drug producers in the country — Organisation of Pharmaceutical Producers of India (OPPI), which represents the interest of big and multinational companies (MNCs), and the Indian Drug Manufacturers' Association (IDMA) — have opposed the latest move of the Government. The industry is facing a challenge on another front too, as the pressure to introduce some changes in the Indian Patents Act of 1970 is mounting. Although many of the provisions of the New Drug Policy of 1986 have not been implemented till now, the prospects of a fresh beginning towards a more healthy drug policy, which will also address itself to the interests of the consumers, have brightened.

The Government is reportedly considering the development of the drug industry through the small scale sector, with a view to providing additional employment opportunities, creating facilities for the development of the bulk drug industry to reduce imports on a gradual scale, and also giving a boost to exports. The much-talked about self-sufficiency in the production of quality drugs, and the simplification of the procedures and policies for the development of the industry, are other important aspects of the review of the existing policy. Import substitution through research and development and the development of competitive technologies for the export market are also likely to the reviewed by the Government.

The inclusion of 21 bulk drugs in category two of the 1987 policy in February, 1990, has been criticised by the industry. Under this category.

MAPE (Maximum Allowable Post-Manufacturing Expenses) of 100 per cent is permitted. Industry feels that this move is in violation of the Kelkar Committee guidelines on inclusion of a drug under the purview of price control, and also goes contrary to the spirit of the drug policy, which tries to optimise implementation by minimising the number of drugs under price control.

According to the industry view, Kelkar had developed five criteria for excluding a drug from price control and they were: (a) drugs not produced in the country, provided their consumption is not significant; (b) drugs with a turnover of less than Rs. 50 lakh; (c) new drugs for which technology or process of manufacture has been developed indigenously, (d) live-saving drugs whose availability is more important than price; and (e) drugs where the domestic production structure is sufficiently competitive to prevent the possibility of overcharging.

Industry argues that some of the drugs (at least five), which have been brought under the purview of price control, have sales turnovers of considerably less than Rs. 50 lakh. Industry sources also point out that with the additional list of drugs subject to control, nearly 75 per cent of the marketed drugs have nor to be sold under statutory prices.

With the latest list, the total number of drugs under price control goes up to 145. This, in turn, is divided into two categories; there will be 21 drugs in category one, while the number of drugs in category two goes up to 124. The MAPE is 100 per cent in category two and 75 per cent in category one. It may be recalled that the DPCO of 1987 had reduced the categories of drugs under price control to two, from the three categories under DPCO 1979, which had attempted to incorporate some of the recommendations of the Hathi Committee.

The 1986 drug policy statement, "Measures for Rationalisation, Quality Control and Growth and Development of the Drugs and Pharmaceuticals Industry", was basically in conformity with the economic policy of liberalisation pursued by the Rajiv Gandhi Government. This was quite soft on the industry, although it did try to take into account some of the issues pertaining to the drug industry raised by the consumer movement. It was argued that the Hathi Committee's prescriptions had failed to increase the production of drugs or ensure the availability of drugs at reasonable prices. Thus the new policy wanted to give price incentives to the producers for increasing production.

However, even the recommendation of the new policy (1986) have not been implemented properly, and some of the important ones have not yet been taken up for consideration. The national drug authority proposed for assessing the needs of the country in the pharmaceuticals sector for instance, has been ignored till now. Besides, the expected production of drugs too have not materialised. As against the targets of Rs. 1,033 crore worth of bulk drugs and Rs. 3,775 crore of formulations for the terminal year of the Seventh Plan (1989-90), the estimated production levels for the year are just Rs. 607 crore and Rs. 3,225 crore respectively.

The need to curtain the innumerable irrational combinations, which have led to such a massive proliferation in the production of formulations (estimated at 60,000), too has not been tackled effectively. Thus, the need to draw up a list of essential drugs will be very high on the agendas in any review of the existing drug policy. It may also be mentioned in passing that even the developed countries to not have such vast lists of formulations as available in this country. The production and marketing of banned and harmful drugs too has been continuing, since the drug manufacturers have successfully sought the intervention of the Courts in this respect.

Against this domestic scenario, the export performance of the Indian drugs industry has been relatively better and the export of bulk drugs especially to the developed countries has caught the attention of the advanced West. These countries are now bringing pressure to bear on India, to change certain provisions of the Indian Patents Act of 1970, so that they can control the drug market more effectively. The export of drugs last year (1989-90) is placed at Rs. 600 crore, which reflects a sharp rise over the earlier year's level of Rs. 467.50 crore (see table). Besides, the export of bulk drugs to the United States alone is estimated at Rs. 200 crore. One of the major reasons why the US government is refusing to remove India from the "Super 301" hit-list, has been the impact that Indian bulk drugs exports have made in the US drugs market. This also indicates he clout of the world pharmaceuticals industry, which is in the hands of just around 30 companies which control a trade of more than $ 200 billion.

Exports of Drugs

Year	Value in Rs. crore
1982-83	111.00
1983-84	162.00
1984-85	217.00
1985-86	265.00
1986-87	222.96
1987-88	289.00
1988-89	467.50
1989-90 *	600.00

* Estimated
(Source: IDMA)

At the international level, recent mergers and acquisitions have made the drug industry highly ologopolistic and due to improve research and development the life of each brand has been reduced considerably. Since the Indian Patents Act allows Indian companies to produce new drugs seven year after its introduction, the MNCs controlling the drug industry are unable to market their brands for a longer period. This explains the reason for the pressure to amend the Patents ACt. which would prohibit the domestic development of the drug for 14 years. This would hamper the efforts of Indian drug manufactures in developing bulk drugs. Dependence on MNCs would continue and the latter would, by virtue to their market clout, be able to overcharge the developing countries, including India. It is heartening to note in this context that the Indian sector of the drug industry is strongly opposing the amendments so the Patents Act.

According to a study prepared by the working group for drugs and pharmaceuticals, it has been estimated that by the end of the Eighth Plan (1994-95), the requirement of bulk drugs would be Rs. 810 crore, while for formulations it would be Rs. 4,896 crore. The investment needed to bring about that order of growth is estimated at Rs. 250 crore — Rs. 150 crore for bulk drugs and Rs. 100 crore for formulations. The total investment in the Plan is estimated at Rs. 400 crore (including the outlay needed for setting up two penicillin plants), which is about 50 per cent of the present investment in the industry.

(Source: *PTI Corporate Release*)

E

EDIBLE OIL

Import

The quantity of edible oils imported during 1990-91 was 89,941 MT and the value of imports was Rr. 62.25 crore.

Production and Requirement

The requirement and production of edible oils for the oil year 1989-90 (November-October) were estimated at 57.72 lakh Mts and 47.22 lakh Mts. respectively.

There was a shortage of almost one million tonnes in the overall supply of edible oil in the country in the oil year 1989-90 (November 1989 to October 1990).

The overall deficit in supply of edible oils as compared to demand was roughly 9.1 lakh tonnes in the 1989-90 oil year. To make good the growing shortfall in edible oil, efforts to enhance production through the oilseeds production programme (OPP) and measures to increase the efficiency of oilseed processing and extraction were being taken.

6.07 lakh tonnes of edible oil of the value of Rs 328.32 crore was imported in 1989-90 oil year. The demand of edible oil in the country in oil year 1990-91 is likely to be around 60 lakh tonnes.

Supplies

The Bombay Oilseeds and Oils Exchange Ltd. has strongly urged the Government authorities to devise appropriate and effective measures to streamline the availability of groundnut oil in the Bombay market.

The board of directors of the Exchange viewed with grave concern the dwindling stocks position following the informal ban imposed by the Gujarat Government on the movement of groundnut oil and other edible oils to other States, including Maharashtra State.

The population of over 10 million in Bombay entirely depends for its consumption requirements on outside supplies, particularly, Gujarat State. As a consequence of the ban, groundnut oil and other edible oils witnessed alarming spurt with groundnut oil having shot up from Rs 251 as on March 26 to Rs 263.50 on March 31, 1990 and to Rs 267.50 on April 2, 1990.

There are disturbing reports in the market that Andhra Pradesh Government is also contemplating to impose restriction on the movement of groundnut oil. The situation is further worsened by the stoppage of supplies by NDDB and the reduction in supplies under the public distribution system.

(Source: *Hindustan Times*)

EDUCATION

See also Operation Black Board; Literacy; Schools in Delhi.

Budget Allocation

Budget provision for education departments of Central and State governments in 1991-91 is about 11.1 per cent of the total revenue budget.

EIGHTH FIVE YEAR PLAN

The Eighth Five Year Plan was to commence from April 1, 1991, covering the period 1992-97. The plan document is yet to be laid on the table of both Houses of Parliament after it has been finalised and considered and endorsed by the National Development Council.

ELECTION

Poll Complaint

The Election Commission received 419 complaints regarding large-scale malpractices during the recent Lok Sabha elections in Bihar.

The Cong-I had made 222 complaints, the Samajwadi Janata party 75, the BJP 60, the Janata Dal 23.

Ballot Papers

About 80 per cent of the used and unused ballot papers and statutory covers for the East Delhi Parliamentary constituency in the recent general elections were reported destroyed in a fire in the office of the Chief Electoral Office, Kashmere Gate.

ELECTRIFICATION OF VILLAGES

As many as 4,73,366 out of over 5.79 lakh villages in the country were electrified as on Nov. 30, 1990. All villages in Haryana (6745), Kerala (1219) and Punjab (12,342) were electrified as on Nov. 30, 1990.

ELECTRONICS

Electronics industry in India has achieved a remarkable growth during the last 3-4 years. During VI Five Year Plan period (1980-85), the industry had grown at a compound rate of 25 per cent. During the VII Plan (1985-90), this growth rate had stepped up significantly; a compound annual growth rate of 35 per cent is likely to be achieved. In monetary terms the production of electronics will increase from Rs. 20,810 million in 1984-85 to Rs. 92,100 million in 1989-90.

Alongwith increase in production, exports are also looking up. Electronics exports during 1985-90 have grown at a cumulative annual rate of 40 per cent compared to 27 per cent achieved during 1980-85. It is estimated that by the end of 1989-90, the exports are likely to cross a figure of Rs. 8,500 million. The export of software has received a special thrust, where exports are estimated to increase from Rs. 300 million in 1985-86 to Rs. 1,750 million in 1989-90.

A review of the performance of India's electronics industry during Seventh Plan in terms of production, pattern of growth, areas of usage, R&D, etc. revealed the following positive aspects:

1. There is an impressive increase in the manufacture of almost all types of electronics items with marked improvement in quality of products available within the country. The prices of certain items like computers have reduced significantly.

2. Investment in electronics sector particularly in components and computer industry has grown substantially. An investment of Rs. 25,000 million has been made in the industry during VII Plan (both public and private).

3. Application of electronics in key sectors of economy particularly process control and industrial electronics has increased. In traditional industries like sugar, textiles etc. use of electronics has enabled improvement in techno-economics of operations.

4. A wide manufacturing base has been established both in small scale sector and large units.

5. A large pool of technical manpower at various levels such as technicians, engineers and managers have been developed.

6. Substantial R&D and electronics application base has been built up in a large number of areas. Carefully chosen projects exploiting this could enable development of contemporary technologies/products.

7. Infrastructure has been built for Standardisation, Testing and Quality Control. Institutions such as, CEDT, ER&DC, etc. have been established.

8. Computers are being used in all sectors of services including business, education and offices. Computerisation of Rail and Airline reservations have already begun and some activities have been initiated in the banking sector.

Licence to Clarify Electronic Products

Three Indian laboratories have been licensed to test and clarify electronic products manufactured in India for international markets for the first time. This facility has been brought in by the Department of Electronics to help Indian manufacturers of electronic goods to get their products certified for international competitiveness.

This has been done under the International Electro-Technical Commission for quality assessment system. These laboratories belong to the Directorate of Standardisation, Testing and Quality Control (STQC).

India has a formal understanding with international laboratories in countries such as West Germany and USA for carrying out safety testing and certification.

At present, consumer electronic products and manufactures of information technology products have been sending their goods to laboratories abroad for certification. The present facilities at the three Indian laboratories will help these manufacturers cut down their expenses and remove uncertainty involved in sending their products to foreign countries for such testing and certification.

The Department of Electronics has also taken steps to certify electronic products from safety angles as well. Safety certification is a mandatory clause for entry into Western markets for any electronic product.

(Source: *Hindustan Times*)

ELEPHANT POPULATION

Twenty-one elephants were reported killed by poachers in Karnataka between 1987-88 and 1989-90.

Steps taken to prevent poaching of elephants and smuggling of sandalwood included a ban on trade in ivory and joint operations by the forest and police departments of Tamil Nadu and Karnataka.

The estimated population of elephants at present is between 15,700 to 17,500. A committee of chief wildlife wardens of the four southern States has been set up to coordinate the antipoaching activities.

EMPLOYEES' STATE INSURANCE (ESI) SCHEME

The percentage of the total labour force including those in small, rural and unorganised sectors covered by the Employees' State Insurance (ESI) Fund health benefits was 2.64.

ENCEPHALITIS

Japanese encephalitis was prevalent among both tribals and non-tribals in rural areas and the incidence was more among the people associated with pig rearing 1291 persons died due to this disease.

ENERGY DEMAND

The energy demand by the end of the Eighth Plan is estimates to be of the order of 385 billion units. The installed capacity in the country as on end-March 1990 is 63,992 MW, consisting of 44,130 MW thermal, 18,297 MW.

ENVIRONMENT POLICY (DRAFT)

A draft on National conservation strategy and policy statement on environment and development has been prepared.

EXCAVATIONS

The excavations of exquisitely carved "four-member" pillar, microdrill bits of hard stone and a huge walled middle town from the ancient site of Dholavira in Kutch have added a new dimension to the Indus Valley Civilisation and its expanse from Mohenjodaro in the North to Sourashtra in the South.

The finds rated as the first of its kind ever excavated from any of the existing sites of the Indus Valley Civilisation are expected to throw new light on the Harappan culture, the mysteries of which are being currently unravelled at Dholavira in Kutch, Gujarat.

The excavations conducted by the Archaeological Survey of India (ASI), from January 24, 1990 have opened up a "magnificent Harappan city" of unique planning and also of enormous accumulation.

The ancient archaeological site of Dholavira that over the years aroused the expectant curiosity of the archaeologists both in India and also outside has now brought into focus a new find that may prove more important than what has been excavated at Mohenjodaro, Harappa and Kalibangan.

Archaeologists are excited over the find of four members of a pillar consisting of an exquisitely carved demaru-like base and three pieces of the shaft tapering upwards. The other important finds include besides the microdrill-bits of hard stone, 12 seals (complete or fragmentary) and six seal impressions on clay and one bronze figuring of an animal.

Many other funds like beads of semiprecious stones, gold, copper, shell, steatite, faience and clay, usual copper objects including a pin with two spiral heads, bangles and stone shell, clay models of cart-frames, wheels, animals, gamesmen, triangular cakes of terracotta and a variety of queens, rubbers, polishers and pestels and mortar used for domestic as well as manufacturing purposes.

The site first discovered in 1967-68 is also expected to help the historians to arrive at a conclusive outline of the land route followed by the Harappans in their migration from the North to the South-East.

The excavations have also revealed an entire city complex, secured by massive mud brick defences, bifacially venered with stones. Inside, there are extensive open spaces surrounding as well as lying in all the three principal divisions — the fortified citadel having two subdivisions, the enwalled middle town and unwalled lower town — all acclaiming almost 48 hectares.

The citadel is enravelling more than 17 metres broad fortification walls furnished with gates of which those on the East and the North are provided with chambers, passage ways and flights of the steps.

Of the middle town, the walls on the East, North and the South besides the structures built inside have thus far been confirmed.

The excavations have also opened up houses with rooms, fireplaces, storage jars, sample or cutstone drains and sullage jars or pits. Some of the structures are reported to have indicated specialised craft activities such as bead-manufacturing, shell working, etc.

4.5 metre thick deposit on the citadel have already been examined while much-accumulation still remains to be unearthed.

In the last four decades more than 250 Harappan sites have been found in the Indo-Pak sub-continent and some of these have also been excavated. The unearthing of the massive finds at Dholavira are considered among the major excavations equalling the ones unearthed at Mohenjodaro, Harappa and Kalibanga. However, the import of the latest finds at Dholavira lies in the fact that the magnificent Harappan cultural centre was the strongest citadel that controlled and protected the eastward movement of the Harappans in the Kutch region.

The excavations at Dholavira also called Kotadi have also established that the Harappans used a land route to reach Kutch and then proceeded to North Gujarat, and Saurashtra from Surkotada after crossing and Little Rann. For trade and other immediate needs, the historians believe that the Harappans might have also made use of the sea route.

On the basis of radio-carbon dating, the 2300 B.C. to 1750 B.C., Indus Valley Civilisation blanketed a vast expanse from Dher Majra in the foot-hills of Siwaliks in Punjab to Malwan on the estuary of Tapti, and from Baragaon district Saharanpur in U.P. to Sutkagendor in the Makran region in Baluchistan varying from matured to degenerated phase of the culture.

(Source: *Hindustan Times*)

EXPORTS

See also Agriculture: Exports; Cashew nut: Export; Coir: Export; Defence: Defence Exports; Knitware: Export; Tea: Export.

Export Contracts

The Government has decided that all export contracts with foreign buyers in general currency area countries shall be in foreign convertible currency only.

Export Decline

As many as 11 major sectors suffered decline in respect of export growth during 1990-91 as compared to 1989-90.

The sectors which suffered decline include plantations, agricultural and allied products, ores and mineral engineering goods, textiles, handicrafts, carpets and petroleum products.

Exports by the State Trading Corporation (STC) fell sharply by Rs 393 crore during 1990-91 as compared to the previous year.

STCs exports during 1990-91 amounted to Rs 359 crore as against Rs 752 crore in 1989-90.

The total turnover of the STC also declined marginally from Rs 1,855 crore in 1989-90 to Rs 1,819 crore in 1990-91.

The import sales, however, in creased from Rs 1,070 crore in 1989-90 to Rs 1,403 crore in 1990-91.

STCs channellised exports also

declined from Rs 135 crore in 1989-90 to Rs 114 crore in 1990-91, the report said.

However, increases over the previous year's export were achieved in respect of sugar, alcohol, molasses and opium.

Non-channelised exports declined from Rs 262 crore in 1989-90 to Rs 167 crore in 1990-91.

Exports to USSR

1991 trade plan envisaged Indian exports to USSR of Rs. 5,081 crore and imports Rs. 4,330 crore.

During the period January 1 to April 30 this year, Indian exports to the USSR this year, Ineian exports to the USSR amounted to Rs. 1,505 crore and imports Rs. 891 crore, he told Shri S Chanpuria.

Engineering Export

According to the 33rd Business Outlook Survey of select 128 engineering exporting companies, conducted by the Confederation of Engineering Industry (CEI).

73 per cent of the responding exporting companies forecast bright export prospects. Twenty-four per cent expect the same trend to continue and only 3 per cent foresee a deterioration in export prospects in the next six months. This was against 61 per cent, 30 per cent and 9 per cent when the last Survey was conducted in October 1989.

Sectorwise break-up of the analysis reveals an overwhelming majority i.e. 88 per cent of the exporting companies forecast optimism with regard to export prospects in the Basic Metal sector, 74 per cent in the Transport Equipment Sector, 69 per cent in the Industrial Machinery Sector, and 63 per cent in the Electrical Machinery Sector forecast a buoyant export climate. Respondents in Metal Products sector, however, expect the present trend to prevail in the near future as reported by 47 per cent of the respondents.

Sixty-one per cent of the exporters expect value of new orders to appreciate, 27 per cent foresee the same trend to continue and 12 per cent apprehend a decline in the value of new orders. This is against 50 per cent, 33 per cent and 17 per cent observed in the last six months respectively.

The value of export deliveries is projected to enhance in the next six months. This is reported by 61 per cent of the exporters, 26 per cent expect no change, while 13 per cent visualise a decline.

Fifty per cent of the respondents to the CEI survey expect no change in the average price of the export order bookings, 33 per cent visualise an increase and 17 per cent foresee a definite decline.

(Source: *Hindustan Times*)

Export of Engineering Goods

(Rs in crores)

Sl. No.	1987-88	1988-89	1989-90
1. East Europe	363.90	450.50	575.90
2. West Europe	173.50	274.50	375.75

Marine and Food Products Export

There has been a substantial increase in the export of marine, fruit and vegetable products during 1989-90. The production of various items in the consumer industry sector and milk products also went up considerably during the year.

The total installed capacity of fruit and vegetable processing industries increased from 5.99 lakh tonnes in December, 1988 to about 7.08 lakh tonnes at December-end of 1989.

The total number of licensed units increased from 3367 on December 31, 1988 to 3629 on December 31, 1989. The exports of fruit and vegetable products increased from 49,389 tonnes valued at Rs 59.89 crore from January to December, 1988 to 63.967 tonnes valued at Rs 79.48 crore from January to December 1989.

In the deep sea fishing sector, the total number of deep sea fishing vessels owned by Indian companies increased from 153 in 1988 to 172 by the end of 1988-89. The export of marine products by December, 1989 was 67,513 tonnes valued at Rs 426.78 crore despite a slump in the major international markets.

Under the scheme of modernisation of rice mills, the number of modernised rice mills had gone up from about 24,500 during 1988-89 to 26.750 in 1989-90.

In the field of international cooperation, two projects in the food processing sector were identified for inclusion under the UNDP country programme. Feasibility reports were being prepared in sectors such as meat and meat products, fisheries, fruits and vegetables. Cooperation possibilities were explored with the USSR, UK, Germany, Australia and France.

(Source: *Hindustan Times*)

Textile Export

Textile exports (excluding jute and handicrafts) including garments are projected to reach a level of Rs 25,000 crore by the year 1994-95 and Rs 50,000 crore by AD 2000.

The revised target for textile exports (excluding jute and handicrafts) is likely to be around Rs 12,102 crore for 1991-92). The target has been revised upwards in the light of recent changes in trade policy and downward adjustment of the rupee.

This represents an increase of around 30 per cent over the actual export performance of Rs 8,410.12 crore in 1990-91. Exports of readymade garments from India amounted to Rs 4,640.64 crore and the target for 1991-92 for garment exports is likely to be Rs 6,960 crore.

India's share of the world trade in garments is estimated to have come down from 15 per cent to only 2 per cent.

India is at present exporting only summer and spring garments whereas there is tremendous potential for winter and autumn garments also. This is a major gap in India's garment exports profile which should be bridged, if necessary, by improving the availability of fabrics through imports.

Fabric imports are already allowed under the Advance Licensing Scheme. However, one difficulty was that input output norms for imports had not been worked out.

The case for differential replenishment rates based on value addition is already under consideration of the Commerce Ministry.

Withdrawal of CCS had put many exporters to financial loss besides affecting exports to non-quota countries. While the excise duties on all inputs going into a garment have been raised in each successive budget, the duty drawback rate for woven and knitted garments has been reduced from 10 per cent last year to 8 per cent early this year.

The recent increase in the rates of interest on pre-shipment and post-shipment credit for exports had been very steep — previous rates were 7.5 per cent and 8.5 per cent for pre and post-shipment credit, respectively.

(Source: *Hindustan Times*)

Volume of Exports of Manufactures

(average annual growth %)

Country	1980-85	1985-88
Hongkong	9.6	14.0
Singapore	5.4	26.1
Taiwan	13.3	28.1
South Korea	13.0	23.9
Malaysia	17.7	13.8
Thailand	15.4	27.8
Philippines	4.5	7.7
Bangladesh	2.6	10.5
Sri Lanka	21.2	2.6
Pakistan	6.4	17.4
Indonesia	146.9	17.2
Nepal	25.9	0.1
INDIA	**0.1**	**12.8**

Source: *India Today*, July 31, 1991.

Textile Export Policy

The Government announced the long term export quota policy for 1991-93 in respect of textiles and garments keeping in view the possible abolition of discriminatory bilateral quotas within a decade or less and unification of the European market in the near future.

The policy endeavours to provide the necessary impetus for the growth and direction of the Indian industry in both domestic and export spheres. The main objectives of the policy are full utilisation of the quotas, maximum realisation of foreign exchange from available quotas and orderly development of the export trade.

According to the new policy, the annual levels of export quotas for garments and knitwear will be distributed among the various systems such as past performance entitlement system - 60 per cent, first come first served system - 10 per cent, manufacturer exporter entitlement - 18 per cent, public sector entitlement system - 2 per cent and non-quota export entitlement system - 2 per cent.

In the case of textiles, including yarn, fabrics and made-ups, the percentage of the allocation against various systems will be past performance entitlement system - 40 per cent, contract reservation entitlement system 20 per cent, first come first served system - 20 per cent, public sector entitlement system - 20 per cent and manufacturer exporter entitlement system - 12 per cent.

The significant changes made in the policy have been to reduce the percentage of allotment under the CRE system to 25 per cent and of FCFS to 20 per cent. Along the lines of garment policy, 3 per cent reservation for public sector units like NTC and State Textile Corporations etc., has been introduced. In order to encourage the modernisation and upgradation of technology, a new system of MEE equal to 12 per cent of the annual level has been introduced for exclusive allotment to manufacturer-exporters who have undertaken substantial investment in modernisation and upgradation of technology in their plant and machinery. The guidelines in this regard will be issued by the Textile Commissioner.

While the policy will be in operation for three years, the Government reserves the right to amend the policy as necessary depending upon the outcome of the Uruguay Round of negotiations and the future of MFA. The Government is hopeful that the new quota policy will give the necessary momentum to the modernisation of the textile industry and lead to increase in the export of higher unit value textile products.

The Government appointed a task force to evolve a new quota policy for garment and textile exports.

The other factors which went into consideration of the new quota policy were the rapid growth of garment and textile exports over the last few years and the tremendous potential in future. The policy seeks to provide for a proper blend of stability for established exporters and opportunity for new - comers. Better facilities for manufacturers including public sector organisations and apex corporations with genuine manufacturing capacities have been provided to give a proper direction to the development of export trade.

(Source: *Hindustan Times*)

EX-SERVICE MEN

The total number of exservicemen yet to be rehabilitated as on January 1, 1990 is 2,59,780 as per the figures in the live registers of persons seeking employment in various States and Union Territories.

EYE DISEASES

Cataract

As many as 18 million people in the country were suffering from cataract. According to a medical survey report 81 per cent of the blindness in the country was due to cataract.

F

FAO TECHNICAL COOPERATION PROGRAMME

The Government of India and the United Nations Food and Agriculture Organisation (FAO) signed two agreements for new projects under the FAO Technical Cooperation programme valued at US $250,000 and $59,000 in the areas of policy analysis using farm level data and wildlife development.

FAIR PRICE SHOPS

There were 3,58,490 fair price shops and public distribution outlets throughout the country on December 31, 1989. The rural areas account for 2,70,446 and the urban areas 88,044.

FAIRS

The Trade Fair Authority of India spent Rs. 873 crore on an exclusive Indian exhibition at Manila and participation in 20 other foreign fairs.

The foreign exchange earned for the country in the form of business booked in these fairs was Rs. 275.76 crore.

FARM SECTOR

The additional income accrual to the agricultural sector during the 1990-91 'kharif' season as a result of the revised pricing for the crops as recommended by the Hanumantha Rao Committee, will be Rs. 3310 crore.

Similarly, for the 'rabi' crop this year, this additional income accrual will be Rs. 795 crore. For the rabi crop, the revised pricing formula was announced on April 20 this year and for kharif crop, the announcement on price increase was made on June 16 this year.

(Source: *Hindustan Times*)

FEDERATION OF INDIAN CHAMBERS OF COMMERCE AND INDUSTRY (FICCI)

FICCI has set up 16 expert committees for the year 1990-91. These committees, set up under the chairmanship of eminent industrialists will provide direction and thrust to FICCI's activities during the year.

The Chairmen of the sub-Committee are: Banking and Finance K.K. Modi; Foreign Trade, G. P. Goenka; Industry, A.K. Rungta; Internal Trade, A.S. Kasliwal; Taxation and Company Law, R.S. Lodha; Agriculture, Rural Development and Social Welfare, V.S. Agarwal; Transport, Kulwant Rai; and Ecology and Environment, Deepak Banker.

In addition to the 8 sub-committees, there will be 8 more ad-hoc committees. Chairmen are: Sudhir Jalan, Energy Committee; Dr. Gurpreet Singh, Management and HRD; Dr S.K. Somaiya, Small-Scale Industry; S.R. Jiwrajka, Electronics and Telecommunication; J.N. Guzder, Consumer Business Forum; Jehangir J. Cama, Tourism; Bansi Dhar, Technology; and H.V. Goenka, Housing and Public Works.

(Source: *Hindustan Times*)

FENCING OF BORDER

Fencing of 356.2 km along the Indo-Pakistan border in Punjab and another 58.8 km in Rajasthan had been completed.

Further fencing of 144 km on the border in Rajasthan had been approved and work was likely to be taken up shortly.

Over 36.5 km on the Indo-Bangladesh border had been fenced. There was also no dispute with Bangladesh on the fencing issue.

Rs. 325,12 lakh have been sanctions for fencing of 40.35 km of Indo-Bengladesh border in Assam.

FERTILISERS

Import

The total value of finished fertilisers imported during 1990-91 is Rs. 133.82 crore and the subsidy given on indigenous and imported fertilisers during the year was Rs. 4389.06 crore.

Subsidy

The amount of subsidy given on fertilisers (including imported) during 1989-90 was Rs. 4,000 crore.

FIRES IN DELHI

During the period from April 1, 1989 to March 31, 1991, 71 incidents of fire took place in Delhi and 43 people lost their lives.

Four persons were killed in a serious fire which broke out in the *jhuggies* near the old Yamuna bridge. The probable cause of the fire was a short circuit and the loss of property estimated at Rs. 19.65 lakh.

197 fires had taken place in J.J. Colonies in Delhi in 1990-91. Action had been initiated to set up nine more fire stations in the Union Territory.

FISCAL DEFICITS

In its recently published report relating to the problems of Indian economy, the World Bank has indicated that "holding real Government consumption" constant even for five years would help the health of Indian economy "significantly". Of course, this would require control of Government wage bill, net purchase of goods and services and subsidies. Government subsidies totalled Rs. 10,677 crore in 1989-90.

The World Bank has recommended a close monitoring of defence expenditure, elimination of projects of questionable economic or social value and timely execution of other projects. Continued use of adjustments like hiking interest rates, portfolio restrictions on financial intermediaries, raising of statutory liquidity ratio and cash reserve ratio were not sustainable solution.

The Bank has suggested flexible management of the exchange rate for diminishing the profitability of imports while enhancing possibility of increasing exports. Other recommendations aim at improving the efficiency of resources mobilisation and its utilisation by public enterprises (PE). It expects that resources mobilisation would play an important role in maintaining macro-economic balance.

Even a three per cent return on the resources by PES would contribute substantially in increasing public sector savings. This would, however, need structural changes, necessitating correction of the incremental capital-output ratios of this sector, which are presently about thrice those of the private sector and solving the "overmaning" problem.

The Bank is all praise for the new practice of execution of a memorandum of understanding (MOU) between the Government and the public sector enterprises.

It is, however, clear that the Indian economy has run into a vast guagmire and unless immediate pragmatic measures, with a will to enforce these, are adopted, a deep crisis of wide dimensions is starting the nation in the face.

The dark clouds: the written-off and bad-debts, including large-scale wilful default camouflaged as "problem recoveries", of the banking system, total over Rs. 44,000 crore. The annual generation of black-money is placed at a minimum of Rs. 40,000 crore, besides an injection of another Rs. 40,000 crore by smugglers — total annual addition of Rs. 80,000 crore of black-money.

The internal debt of the Government now totals over Rs. 2,38,000 crore and its outstanding liabilities total Rs. 2,68,000 crore, which is 60 per cent of the gross domestic product. Of the total outstanding liabilities, internal debt constitutes 30 per cent of GDP.

But the darker cloud is the external debt and what makes it more fearsome is the daily deteriorating balance of payments position. India's external debt liability is more than Rs. 80,132 crore. No amount of ostrichian attitude can tide over the situation: at current prices, external debt is over 18 per cent of the GDP. Still worse is that the debt export ratio has crossed 220 per cent mark and worst is that debt-service ratio of debt-service payments exclusive of interest on non-resident (NRI) deposits — to current receipts (exports plus invisible receipts), has jumped up from eight per cent to 21.6 per cent over the last one decade. This is Government of India's figure on March 1990. The World Bank, however, has placed it at 26.4 per cent.

This is a disquieting signal. 30 per cent is the "critical point" and India is nearing it. It must be noted with deep concern that India's external debt figure has been climbing the graph fast.

(Source: Harish Bhat, *Hindustan Times*)

See also Government Budget Deficit.

FISH

The estimated fishery potential of the Indian exclusive economic zone which embraced the Arabian Sea, Indian Ocean and Bay of Bengal was 3.900 million tonnes. The quantity of marine fish landed in the country during 1990-91 was about 2.3 million tonnes.

See also Prawn export.

FISHERMEN

Nearly 300 fishermen and their boats were captured on March 27 from the high seas by armed groups of men and detained at a point on the northern Sri Lankan coast. The Fishermen and their boats were released on March 30, and returned to India.

FIVE YEAR PLANS

See Eighth Five Year Plan.

FLOODS

Control Measures

Of the 219.76 lakh hectares of land liable to be affected by floods of the Ganga river in the basin States, a total of 66.53 lakh hectares had been afforded a reasonable degree of protection till March 1990.

In the sub-basin States of the Ganga, physical flood control measures only could not offer protection against all floods which would continue to inflict loss and damages.

In this context, flood forecasting activity assumed greater importance for organising flood fighting and mobilising relief. Flood forecasting network on all major rivers of the Ganga sub-basin had already been established by the Central Water Commission.

FOODGRAINS

Stock

The foodgrain stock were 21.07 million tonnes with public agencies. The position as on July 1, 1989 and July 1, 1990 was 13.05 million tonnes and 20.31 million tonnes, respectively.

Stocks with the Food Corporation of India (FCI) as on July 1 were 19.8 million tonnes, including 5.57 million tonnes wheat and 8.99 million tonnes rise.

Total disbursement received under the second foodgrains storage project from the World Bank was utilised for the creation of 35.26 lakh tonnes of storage capacity at 211 centres.

FOREIGN COLLABORATION

A total of 465 foreign collaboration approvals were granted between December 1 and July 31, 1990.

A total of 359 foreign collaboration proposals were approved by the Government from January to June 1990.

See also Foreign Investment

FOREIGN CONTRIBUTIONS

Forty organisations were registered under the Foreign Contribution (Regulation) Act, 1976 in Jammu and Kashmir for receiving foreign contributions. Of these 17 organisation received Rs. 1.77 crore in 1990.

FOREIGN EXCHANGE REGULATION ACT (FERA)

A strategy paper prepared by the PHD Chamber of Commerce and Industry (PHDCCI) has suggested that the Government should substantially liberalise the Foreign Exchange Regulation Act (FERA) towards making the overall system more responsive to the changing needs of the European Economic Community (EEC) markets.

The paper says that it would help the country in further consolidating its position in the existing export market and it would also help in the promotion of new products.

The Chamber proposes to act as a catalyst for creating greater awareness amongst its members about the emerging market of EEC and recommended for the setting-up of a full time EEC desk in order to maintain a constant interaction with various agencies in EEC and with our embassies. This would help the country in getting more accurate and timely information which could also be disseminated among its member so that they could avail the maximum benefit. It would also organise seminars and workshops.

In order to offer successful competition, the exporters would have to continuously upgrade their quality and productivity. This could be attained only through proper research and development work.

The competitiveness is a total concept and is not restricted merely to price and quality. In the develop and sophisticated markets like EEC quality is certainly a very important aspect. The other important factors which could also affect the export performance includes timely delivery, effective communication, product design and conforming to the fast changing consumers' preferences.

Indian exporters should spend considerable amount on regular market research and survey on their line of products and adapt themselves to the changing scenario in time.

Our export effort is quite diluted in view of the presence of a large number of small exporters. The sophisticated and a large sized market like that of EEC would necessitate consolidation of efforts. It is felt that a consortium approach is extremely necessary. Similar product line exporters should form cooperative or a consortium like body. This would give them a better bargaining power.

Every exporter should build-up a reliability and lasting relationship with a select group of buyers and in select products.

(Source: *Hindustan Times*)

FOREIGN INVESTMENT

Total investment by foreign parties in Central Public sector enterprises as on March 31, 1990, was Rs. 140.27 crore equity capital and Rs. 14,181.28 crore loans, including deferred credits.

See also Foreign Collaboration.

FOREIGN TOURISTS

The foreign tourist traffic to India excluding the nationals of Pakistan and Bangladesh during the first 11 months of 1990 had shown a decline of 0.2 per cent as compared to the same period in the last year.

A total of 11,81,978 foreign tourists have visited India during 1990 (up to November) as against 11,82,832 in the same period in 1989. Country earned an estimated foreign exchange of Rs. 2,265 crore in 1990 as against Rs. 2,006 crore in 1989.

In growth rates of 8 per cent in foreign tourist arrivals and 18 per cent in foreign exchange earnings from tourism were envisaged for 1990-91. Year 1991 had also been declared as "India Tourism Year" to improve the image of the country as a holiday destination.

Several steps including liberalisation of charter policy and introduction of multi-entry visas for a period of five Years, were initiated by the Government to give a thrust to tourism in India.

The Government earned Rs. 2,444 crore in foreign exchange from tourists during 1990-91.

FOUNDATION LAYING CEREMONIES AT DELHI

Ten foundation-stone laying ceremonies took place in the Capital from November 1990 to May 1991.

Among the projects for which foundation-stones were laid but financial approval by the competent authority was still awaited were: The Rs. 15 crore, 100-bed hospital at Raghubir Nagar, the Rs. 11 crore Lok Nayak Setu over the Yamuna near the ITO, and the Rs. 3.07 crore second Inter-state bus terminal at Anand Vihar in East Delhi, Mr Chavan said.

Projects for which the New Delhi Municipal Committee (NDMC) had given expenditure sanction but were still awaiting the Lt-Governor's approval were: the Rs. 4.01 crore project to construct a sub-district-cum-community centre at Gole Market, the Rs. 0.46 crore home for aged persons at Netaji Nagar and Rs. 1 crore project to build a subway at the inter-section of Kasturba Gandhi Marg with Connaught Place.

The project which had obtained all the necessary clearances and on which work was going ahead were: the Rs. 0.22 crore guest house at Chitrakut, the Rs. 330,950 project to develop a children's park at Vonoba Puri in Lajpat Nagar, and Rs. 298.02 lakh project to build an MCD office at Rohini, and the Rs. 17,503,431 Narela housing project.

FRAUDS IN BANKS

As many as 5,312 cases of fraud in the public sector banks were reported from 1988 to 1991, involving over Rs. 198 crore.

FRENCH NATIONALS IN PONDICHERRY

There were 14,267 French nationals in Pondicherry. Indian Constitution does not recognise the concept of dual citizenship.

FRUITS AND VEGETABLES

Production of fruits and vegetables in the country has registered a five-fold increase during the last 20 years. It was 2.40 lakh tonnes in 1989 compared to 52,274 tonnes in 1970. The number of licences to fruit and vegetable processing units increased to 3,629 this year against 1,994 in 1970. There has also been significant increase in the exports of fruits and vegetables. Rs. 79.48 crore worth of fruits and vegetables were exported in 1989 against Rs. 21.90 crores in 1980.

(Source : *Yojana*, December 15, 1990)

G

GANGA ACTION PLAN

The Ganga Action Plan, launched in 1985, failed to complete 192 of the 262 sanctioned schemes within the target till June 1989, according to the Comptroller and Auditor General of India (CAG).

In respect of 23 schemes, approved in the Rs. 196.23 crore Phase-I, even preparatory work had not started.

Of the international assistance of Rs. 64.24 crore, only Rs. 11.48 crore could be utilised.

An amount of Rs. 138.43 crore had been spend on the Ganga Action Plan till June 1989.

(Source: *UNI Release*)

All schemes under the Ganga Action Plan are likely to be completed by December 1993. 261 schemes had been sanctioned under the action plan for which the revised outlay was Rs. 356.72 crore. Out of these, 172 schemes had been completed till June, 1991 with an expenditure of Rs. 245.02 crore.

GEOLOGICAL MAPPING

See Mines and Mining : Geological Mapping

GIRL CHILD LABOUR

The number of girl child labour in the country is estimated to be about 7.6 million according to the latest sample survey conducted by the National Sample Survey Organisation.

See also Child Labour; Children; Women

GOLD

Holdings

As per RBI report the gold holding of the bank as on August 9, 1990 was 332,562,924.09330 fine grams. There is an increase of 7,574,093.52582 fine grams in the gold holding of the bank during the last six months.

Prices

Standard gold

(Rs. per 10 gm at Bombay Bullion Market)

Date	Price
March 4, 1991	3,528
May 14, 1991	3,830
June 29, 1991	3,732
July 2, 1991	3,924
July 3, 1991	4,380
July 5, 1991	4,150

Source: *India Today*, July 31, 1991.

Seizure

During the financial year 1990-91, (April to November 1990) gold worth Rs. 118.24 crore, to totalling 3,468 kg was seized.

Seizure Sold

A total of 9.799 tonnes of seized old was sold during the last three financial years.

No seized gold was sold in 1988-89, 5.630 tonnes and 4.169 tonnes were respectively sold in 1989-90 and 1990-91.

GOVERNMENT BORROWINGS

See also Fiscal Deficit.

Government Borrowings

(Rs. in crores)

Year	Amount
1981 below	10,000
1989	30,000
1990	38,475
1991	43,331

Internal Borrowings

The net internal market borrowings of the Central Government outstanding on March 31 this year were Rs. 70,565 crore. They were Rs. 62,565 crore and Rs. 55,161 crore in the corresponding periods of 1990 and 1989.

The major owners of Central Government securities were RBI (20.1 per cent), scheduled commercial banks (52.2 per cent) and LIC 912.5 per cent).

External Debt

India's external debt outstanding on March 31, 1990 was Rs. 79,982 crore. This was 15 per cent higher over that of Rs. 69,361 crore as on March 31, 1989. IN dollar terms the increase is only about six per cent.

India had an outstanding liability of Rs. 66043.72 crore on external loans as on March 31, 1991.

Total outstanding foreign debt (excluding NRI deposits)

(Rs. on crores)

Year	Amount
1984	27,673
1988	54,650
1990	80,000
1991*	100,00

Source: *India Today*, July 31, 1991.
Note:
* Based on estimates before the devaluation; NRI deposits add up to Rs. 20,000 crore.

Internal Debt

Internal debt which was Rs. 96,505 crore at the beginning of the seventh plan (1985-86) rose to Rs. 2,38,096 crore at the end of the plan (1989-90), while external debt (at the prevailing exchange rate) rose from Rs. 35,725 crore to Rs. 80,132 crore between these two periods. The debt service ratio in 1989-90 of external debt as a proportion to exports and gross invisible earnings in estimated at 21.6 per cent.

Government Internal Debt
Total Internal Debt

(Rs. on crores)

Year	Amount
1981	30,864
1985	60,000
1998	100,000
1991	160,000
1992*	171,635

* Interim Budget Projection.
Source: *India Today*, July 31, 1991.

Cutstanding Debt

The total amount of domestic and external debt outstanding against India as on March 31, this year is estimated to be Rs. 2,38,096 crore and Rs. 80,132 crore, respectively. The amount of interest payment during 1989-90 on domestic and external debts is estimated to be Rs. 16,226 crore and Rs. 3,559 crore, respectively.

External Loans

See External Debt

GOVERNMENT BUDGET DEFICIT

See also Fiscal Deficits.

Budget Deficit

(Rs. in crores)

Year	Amount
1981 below	5,000
1986	5,000
1990	10,000
1992*	9,977

* Net RBI credit to Government, i.e., printed money.

Budget Deficit (as estimated on January 11, 1991)

The actual Budget deficit on November 30, 1990 — at the end of the first eight months of the 1990-91 financial year was Rs. 13,000 crore. At this time, in the preceding year 1989, the deficit was Rs. 13,082 crore.

The current year's deficit so far represents a disquieting picture of shortfalls in revenue receipts, while expenditure continued to be incurred as budgeted. There has been under-funding of the requirements for the Farm Loan Waiver Scheme.

The Gulf crisis also led to a deterioration in the fiscal situation because the Oil Co-ordination Committee (OCC) surpluses would not materialise. The burden on account of fertiliser subsidies would increase and the cost of repatriation of Indian citizens from Kuwait would have to be borne by the exchequer. There will be a shortfall in revenue attributable to the squeeze on non-oil imports and contraction of output.

As announced on December 27, 1990, action has been taken to mobilise additional revenue estimated at Rs. 1,200 crore during the current financial year. A strict rein on expenditure was being maintained.

In the second batch of supplementary demands presented to Parliament on december 27, 1990 the net cash outgo will be only Rs. 649 crore. This represents the total net additional outgo on account of supplementary grants so far during the current financial year. Last year, at this stage, three supplementaries aggregating to a net cash outgo of Rs. 2,119 crore had been presented.

The combination of measures contain the fiscal deficit of the Government at 8.3 per cent of Gross National Product (GNP) in 1990-91. The processes of correcting the serious fiscal imbalance has begun. Consolidation will be effected in 1991-92.

Receipt and Expenditure of Union Government

	1990		1989	
	April-November (in Rs. crores)	% of Budget Estimates 1990-91	April-November (in Rs. crores)	% of Budget Estimates 1989-90
		Tax Receipts		
Corporation Tax	1001	16.44	1233	25.93
Income Tax	2233	41.16	2299	54.15
Customs Duties	12468	58.10	11192	62.60
Union Excise Duties	14859	59.14	13683	60.27
		Capital Receipts		
Marking Borrowing	5653	70.60	3135	42.40
Special Deposits	2382	32.80	3488	59.60
Net External Assistance	1204	27.80	953	24.30
Net Small Savings Collections	3642	72.84	3865	84.02
		Expenditure		
Major Subsidies	5748	67.50	4994	66.84
Write off of Farm Loans	1000	100.00	Nil	Nil
Interest Payments	8532	41.00	8428	49.60
Central Assistance to States for State Plans	6394	60.70	5024	65.70
States Share of Taxes, Duties & Revenue deficit grants	10024	59.10	8631	56.90
Loans to State Government against small savings collections	4647	103.26	4076	99.40

Note: Figures are provisional.

(Source: *Hindustan Times*)

GOVERNMENT EXPENDITURE
See also Non-Plan Expenditure.

Central Government Expenditure
(Rs. in crores)

Year	Amount
1981	20,000
1985	40,000
1987	60,000
1989	80,000
1991	100,00
1992*	110,167

* estimated
Note: Year ending March.

GOVERNMENT FLATS
See also Allotment of Government Residential Accommodation.

Unauthorised Occupations

Several instances of unauthorised occupation of government accommodation which remained undetected for years by the Directorate of Estates have been pointed out by the Comptroller and Auditor General of India.

In his report presented to Parliament recently, the CAG said an official in occupation of a government accommodation in the general pool retired from service in April 1980. The fact of his retirement was noticed by the Directorate only in May 1986.

Unauthorised occupation of accommodation thus remained undetected for about six years. Eviction proceedings were initiated in July 1986 and the house was vacated in February 1987. Huge amount was pending as the whereabouts of the official were reportedly not known to the Directorate.

As official in occupation of a government flat was transferred to an office in Delhi Cantonment in January 1982, due to which he became ineligible was noticed by the allotment section of the directorate only in March 1987, when the rent section intimated non-receipt of licence fee.

Action for cancellation of allotment with retrospective effect was initiated by the directorate in July 1988. The accommodation continued to be under unauthorised occupation (July 1989).

As official in occupation of a Government flat and transferred to a station outside Delhi in October 1982, expired in April 1985. The transfer of the official was noticed by the Directorate only in August 1985 when the deceased official's son applied for regularisation of the said flat in his name on compassionate grounds.

A test check of registers for 1987-89 to watch recovery of damages from unauthorised occupants revealed delay in cancellation of allotments in 734 cases. In 26 cases, the delay was more than 60 months.

Test checks of registers for the same period revealed that 4,292 quarters remained under unauthorised occupation for varying periods, of which 2,553 quarters were got vacated during 1987-89 leaving 1,739 quarters still under unauthorised occupation as on March 31, 1989.

In 116 cases, the period of unauthorised occupation ranged from 60 to 165 months as on March 31, 1989. In 2,553 cases, the period of unauthorised occupation ranged up to 196 months.

(Source: *Hindustan Times*)

GOVERNMENT FOREIGN DEBT
See also Government Borrowings.

GREAT BUSTARD

Present estimates of the population of the Great Bustard in India, based on surveys done by the Bombay Natural History Society, vary between 770 and 1920, the society had in December 1989, sent a proposal to the Gujarat Government for declaring the grassland areas between Naliya, Jakhan Port, Suthri village and Kothara in Kutch district as a sanctuary for the bird.

GROSS NATIONAL PRODUCT

India has suddenly forged ahead of China in per capita GNP.

Chine's per capita GNP is $ 330 and India's is US $ 340 for fiscal years 1988-90.

Earlier, China's GNP per capita for 1988 was $ 340 against $ 310 for the years 1986 and 1987.

India's per capita GNP of 340 is unchanged from the 1988 figure. India's per capita GNP during 1987 was $ 310 — same as China's that year.

Pakistan, which is in the Middle East division, remains ahead of India with $ 350 per capita.

Sri Lanka and Maldives are ahead of both India and Pakistan. Sri Lanka's GNP per capita during 1988-90 is given as $ 420 and that of the Maldives at $ 410, Bhutan's per capita GNP was $ 180. Bangladesh is the poorest in South Asia, with a per capita of only $ 170.

Among other Asian countries, South Korea's GNP per capita was $ 3,600, Malaysia's $ 1,940, Fiji's $ 1,520, Thailand's $ 1,000, the Philippines $ 630 and Indonesia's $ 440.

(Source: *Annual Report, 1990*, World Bank, Washington D.C. reported in *PTI Release*)

GROWTH CENTRES

The Comptroller and Auditor General of India has criticised the slow progress in the coverage of no-industry districts for Central assistance for development of infrastructure facilities.

53 per cent of the identified *no-industry districts* had remained uncovered, through the scheme had been in operation for years. Central assistance was released only to 35 out of 51 approved growth centres. None of the 13 growth centres for which first instalment of assistance was released by the Ministry during 1985-86, was completed till March 1989. Despite liberalisation of the scheme in 1986 in respect of North-Eastern States, the scheme made little progress in that region as Central assistance was released for only one growth centre.

Inadmissible and irregular payments of transport subsidy aggregating to Rs. 1.64 crore were noticed during test-check in audit. Irregularities included payment for transportation of material not covered under the scheme, excess payment due to reckoning rates higher than admissible, payment made to ineligible unit and payments made without verification of documents. The industrial units in the Union Territories of Andaman and Nicobar Islands were not paid transport subsidy during 1982-89, due to delayed fixation of freight rates in June 1987.

Central investment subsidy accounted for about 93 per cent of total expenditure of Rs. 816.75 crore provided during 1980-89. The scheme was extended from year to year from 1975-76 till March 1987 and thereafter in shorter spells till September 1988, imparting an element of uncertainty to entrepreneurs. The Inter-Ministerial Committee, set up for formulating revised incentive scheme, recommended in December 1986, continuance of the scheme with corrective measures during the Eighth Plan period.

H

HIV INFECTION TESTING CENTRE

Testing of blood for HIV infection was started at Christian Medical College, Vellore, Madras Medical College and AIIMS. All the 42 surveillance centres and five referral centres are being linked with blood banks for testing of blood donors for HIV, 28 exclusive zonal blood testing centres had been set up in the four metropolises.

See also Acquired Immuno Deficiency Syndrome (AIDS)

HANDICAPPED

Twelve million people in the country suffer from physical disability of one type or another. Government had earmarked Rs. 32.23 crore for the welfare of physically handicapped persons during 1990-91.

HANDLOOMS

A total of 22 items were reserved for exclusive production on handlooms vide notification dated August 4, 1986.

HAKSAR PANEL REPORT

The Committee was set up in March 1988 to review the performance of the four autonomous institutions of the Department of Culture. These institutions are Sahitya Akademi, Sangeet Akademy, Lalit Kala Akademi and National School of Drama.

The Committee has recommended that chairpersons of the three national Akademies and the National School of Drama should be persons of eminence.

It has said that the primary objective of these institutions should be to promote excellence in fine arts and literature and help the process of conserving and disseminating our cultural heritage.

It has recommended a higher allocations of funds for these institutions and a greater degree of autonomy to the cultural institutions in utilising the available resources.

While making specific suggestions in the case of the four institutions, the Committee has not ignored the common thread binding them. So it has recommended that the national akademies should have much "closer interaction" with their counterparts in various states and must actively involve other cultural organisations and universities in their activities.

It has said the functioning of the corresponding Akademies and equivalent institutions in the States and Union Territories need to be improved. They must be free from political and departmental interference and must be funded and administered properly.

The report says that the akademies, rather than the departments, should award research and promotional grants to institutions and individuals.

While recommending a few measures to improve the functioning of the general councils of the Akademies, the Committee has also noted that it was necessary to link education with culture. It has called for providing basic infrastructural facilities to promote culture in even small towns.

Stating that the zonal cultural centres are functioning without close links with the National Akademies and their counterparts in the States and Union Territories, the Committee says they should set aside a substantial portion of their resources for the creation of infrastructural facilities.

The report says that the status and pay scales of the secretaries of the National Akademies should be identical and equal to that of a Professor in universities.

It has recommended a series of specific measures for the four institutions.

Besides Mr Haksar, the Committee has its members Mr E. Elkaji, Prof Gulam Sheikh, Dr. Prem Lata Sharma, Prof Indira Parthasarathy, Mr H.Y. Sharda Prasad, Mr K. V. Ramanathan and Mr M.V. Ramakrishnan as member secretaries.

(Source: *Hindustan Times*)

HIDDEN SUBSIDY

The Comptroller and Auditor General of India has charged that the callousness in ensuring proper guarantees for external loans has led the Government to pay hidden subsidy amounting to Rs. 718 crore.

There was no centralised control exercised by the Finance Ministry to watch that the loans and interest were paid as per the terms of agreements. Other ministries which had stood guarantee also did not intimate whether any such monitoring was done. Although the Finance Ministry had issued instructions for a consolidated review of all Government guarantees, it was observed that even the Department of Economic Affairs did not undertake such a review.

The amount of guarantees shown in different budget documents for the same year did not mutually tally and there were several discrepancies. The payment of Rs. 17 crore made on account of invoking of guarantee in the 1987-89 by creditors was not reported in budget documents.

The amount of guarantees given by the Government for raising external loans increased by 81 per cent between 1984 and 1988 and stood at Rs. 6,850 crore. Audit scrutiny however showed that the amount as per the budget documents was not correct as it did not include the amount of accrued interest, where the loans were interest bearing and the Government had given guarantee for principal as well as interest. The amount also did not include the guarantee given by the Ministry of Railways.

The amount of guarantee had not been adjusted for fluctuations in the rate of exchange. There was no regular procedure in converting the amount of guarantee into Indian rupees — different ministries followed different procedures and even the same ministry followed different procedures in respect of amount of guarantees provided in favour of different institutions. Where historical rate of conversion was adopted, the actual liability in terms of rupees was much more due to the depreciation of the Indian currency.

No provision existed in the agreements with the institutions for recovery of interest/penalty for delay in payment of guarantee fee. Internal and external loans raised by public sector undertakings, private companies, institutions, etc. are guaranteed by the Central Government for repayment of principal and payment of interest. These guarantees constitute contingent liabilities of the Government.

(Source: *Hindustan Times*)

HIGHWAYS

Funds allocated for maintenance of national highways during 1990-91 was Rs. 15,312 lakh, while in 1989-90 it was Rs. 14,355 lakh.

HINDUSTAN AERONAUTICS LIMITED

The Hindustan Aeronautics Ltd., Ozar, Nasik achieved the production target during 1990-91.

HINDUSTAN FERTILISER CORPORATION LIMITED (HFC)

HFC tops the list of five Public Sector fertiliser undertakings incurring losses.

HINDUSTAN PETROLEUM CORPORATION LIMITED (HPCL)

HPCL has earned a record net profit of Rs. 200.26 crores during 1989-90. The Corporation has been consistently achieving higher profit levels during the 7the Five Year Plan. In 1985-86, the net profit was Rs. 23.33 crores.

Bombay and Visakh Refineries together recorded the highest ever crude throughput of 10.03 million tonnes during 1989-90 and achieved a combined capacity utilisation of 100.3 per cent.

(Source: *Yojana*, December 15, 1990)

HOLY TREE

The holy *'ber' (Ziziphus jujuba)*, popularly known as *Beri Baba Buddha Sahib Tree*, in the *'Parikrama'* of the Golden Temple Amritsar, is perhaps the oldest tree in the world. Located on the northern pavement of the holy tank around the temple, it got its name from the Sikh Saint Baba Buddha. It is believed that the Baba, sitting under the tree supervised the work of the labourers constructing the temple, 400 years ago.

HORSE IMPORT

India imported horses worth Rs. 31.59 lakh in 1988-89 and in the first 10 months of 1989-90, horse imports entailed Rs. 5.35 lakh.

HOSPITALS AND DISPENSARIES

The Safdarjung Hospital here incurred a wasteful expenditure of Rs. 12.70 lakh on setting up a drug de-addiction ward at the hospital which however, could not benefit addicts, according to the latest report of the Comptroller and Auditor General of India (CAG).

It was decided to set up a drug de-addiction ward in Safdarjung Hospital as the treatment of such cases was not found possible in general wards.

An expenditure of Rs. 9.79 lakh was incurred during 1987-88 on the ward which was to start functioning in the hospital by October 1986.

The hospital did not furnish information regarding the installation of equipment and the number of patients admitted in the ward during 1987-89.

The hospital incurred a further expenditure of Rs. 2.91 lakh on staff during 1989-90.

The Health and Family Welfare Ministry in July 1989 ordered the establishment of a de-addiction ward at the Lady Hardinge Medical College and Srimati Sucheta Kripalani Hospital. 11 members of the staff, ordered to be transferred to the ward in the college in September, 1989, did not join and only a limited in-patient service was being provided by the college for drug addicts.

(Source: *UNI Release*)

See also Drug Addicts.

HOUSING

See also Allotment of Government Residential Accommodation; Government Fists.

HOUSING POLICY

The thrust of the National Housing Policy, still in draft form, is more pragmatic than the 1988 NHP which had made the utopian promise of ending "houselessness" by the turn of the century. There is also no talk of housing as a fundamental right, as was rumoured earlier. The new policy has addressed the problem of an ever increasing number of homeless — 30 million families — by a multi-pronged approach. While it is determined to reverse the declining trend in housing investment to total investment (34 per cent in the First Plan to 9 per cent in the Seventh Plan), it lays emphasis on the private investor and on self-construction in building activity. The official agencies will concentrate on making available land complete with basic infrastructure to the public at reasonable rates. There are also various incentive schemes including the use of black money bonds/debentures towards housing for the poor, mortgage schemes and encouraging these by providing insurance cover to the lender against non-payment by the borrower. Other issues the new policy seeks to tackle are to set housing norms and discourage luxury housing by withdrawing some concessions.

At present not even a million houses are being constructed in a year. This is resulting in a build-up of backlog. Not only this backlog but the increasing demand because of population growth has also to be met. There is also the question of providing a minimum standard which has to take into account the increasing number of those who live under the open sky — the pavement dwellers — six million officially.

(Source: *Hindustan Times*)

HYDEL POWER

The total installed capacity of hydro power stations in the country as on March 31, 1991 was 18,752.617 MW. A total 38 hydro schemes with an aggregate installed capacity of about 7,331.6 MW have been cleared by Central Electricity Authority but were awaiting sanction of the Planning Commission.

IMAGE-INDIA

Area

The area of India as on January 1, 1990, is 32,87,263 sq. km. This latest provisional figure reflects more precise assessment resulting from improved techniques of survey and inclusion of certain gulfs and bays as inland waters.

Basic Information

	1980-81	1990-91	2000-01
Population (million)	690	845	1,030
Income per head (Rs.) (at current prices)	1,770	5,540	17,500
Wholesale Prices	100	210	440
Industrial Production	100	207	350/400
Exports (Rs crore)	6,700	35,000	150,000/ 200,000
Imports (Rs crore)	12,500	45,000	150,000/ 200,000
Rupee per Dollar	7.9	18	35
Rupee per Pound Sterling	18.5	35	60

Source: Dubashi, Jay, India in 90's: from frying pair into fire? *Probe India* 11(11) January, 1991, p.63

IMPORTS

See also Coir Import; Crude Oil; Imports.

INCOME-TAX DUES

Advocates/Doctors

Mr Shanti Bhushan and Dr Shiela Mehra are among the highest income tax paying advocates and doctors in Delhi, respectively. Mr J.B. Dadachanji, the Additional Solicitor General, Mr Kapil Sibal, the former Law Minister, Mr A.K. Sen, the former Additional Solicitor General, Mr G. Ramaswamy and Mr O.P. Kahitan figured in the list of top 20 income-tax paying advocates in the Supreme Court.

The list of top 20 income-tax paying doctors included Dr (Mrs) J.N. Bhutani, Dr Hari Mohan, Dr Anand Prakash, Dr Naresh Trehan and Dr S. Padmawati. This had been done on the basis of

Nominal Tariff Rates, 1985

(Per cent of value)

Country	Intermediate Goods	Capital Goods	Consumer Goods	Manufactures Goods
Hungary	14.2	15.0	22.6	20.9
Yugoslavia	18.0	20.7	20.0	19.0
Argentina	21.2	25.0	21.9	22.9
Morocco	21.6	18.1	43.0	27.3
Philippines	21.8	24.5	39.0	28.0
Mexico	25.5	23.5	32.2	24.7
Thailand	27.8	24.8	8.5	33.6
Turkey	29.4	54.9	55.3	37.1
Pakistan	75.0	73.8	127.3	89.8
China	78.9	62.5	130.7	91.2
Bangladesh	97.9	80.5	116.1	100.8
India	**146.4**	**107.3**	**140.9**	**137.7**

Source : *World Bank/The Economist* quoted in *India Today* July, 31, 1991, p 30

their last assessed or returned income.

Out of the 20 cases of advocates, no demand was outstanding in 11 cases and the demand had not fallen due in two cases. In the list of doctors, no demand was outstanding in 12 cases.

Film Stars

Amitabh Bachchan, Sridevi and former Andhra Pradesh Chief Minister N.T. Ramarao are among 41 film stars who owe more than Rs 1 lakh as income-tax.

ITC Limited

Income-tax assessment for the ITC Limited for the assessment year 1984-85 has been completed on July 11, 1990, raising an income-tax demand of Rs 2.63 crore with an interest amounting to Rs 1.27 crore. Income-tax has been paid by the assessee on August 8, 1990. However, on the amount of Rs 1.27 crore by way of interest, the company has filed a waiver petition, which is under consideration.

INCOME-TAX RETURNS

The number of income-tax returns having income of Rs one lakh and above increased from 1.71 lakh in 1988-89 to 2.32 lakh in 1989-90 and 2.38 lakh (provisional) in 1990-91.

INDIAN AIR FORCE

The Comptroller and Auditor General of India (CAG) has taken strong note of the premature failure of many engines of the IL-76 aircraft acquired by the Indian Air Force for meeting its heavy transport requirements. Although need for a fleet of 20 such aircraft had been planned for 1985-90, no cost benefit evaluation was done for setting up indigenous maintenance facilities. The expenditure on repair abroad of 10 engines prematurely withdrawn for overhaul was Rs 5.71 crores. Ten more engines have further failed which require urgent repairs.

The contract for supply of three aircraft to be delivered during 1988 at a cost of Rs 43.50 crore per aircraft and three engines to be delivered within nine to 12 months from the issue of the letter of credit was concluded only in December 1987. While the difference in the costs of engines procured under these contracts was within the norms of escalation as accepted in May 1986, the escalation charged for deliveries of aircraft during 1988 under the contract of December 1987 was 7.77 per cent as compared to the contract of June 1987.

The Ministry had stated that though, as per the protocol, the contract was to be signed by April 1987, the draft offer for the aircraft to be delivered in 1988 was received only in June 1987 and the contract was concluded in December 1987. It added that the escalation factor was 3.6 percent and the increased price was due to variations in exchange rate. Clearly, deviations from the protocol were either not anticipated or, if anticipated, no method had been devised to cater for them.

The prescribed overhaul period of the airframe was 5,000 flying hours whereas that of the engine was 2000 hours or 10 years. Ten engines installed on these aircraft however, were withdrawn prematurely after completing only 9.05 to 38.15 per cent of their prescribed overhaul life. The high rate of premature withdrawals of the engines resulted in low serviceability of the aircraft and consequent low utilisation. As against the sanctioned utilisation rate of 66 hours per aircraft per month, the rate achieved by these aircraft during the period 1985 to 1988 was as low as 15.35 to 28.26 hours.

As a consequence of the exceptionally high rate of premature engine failures, the IAF had to procure three additional engines at a cost of Rs 6.36 crore in December 1987 to maintain the operational status of the fleet since nine out of 10 spare engines procured under the contract of December 1983 and October 1984 had already been utilised to replace the engines withdrawn prematurely. Thus the AIF had to incur an additional expenditure of Rs 6.36 crore on the procurement of three spare engines which otherwise would not have been required.

Also, rotables were withdrawn much before completing their prescribed overhaul life. During the period between August and November 1987 there had been 16 withdrawals of constant speed drive(CSD) turbine units between 500 and 700 flying hours as against the prescribed overhaul life of 2,000 hours. As a result, 22 CSD turbines had to be procured at a higher rate on aircraft on ground (AOG) priority. Also, due to the non-availability of the CSD turbine units two aircraft were grounded during 1987 affecting the serviceability of the fleet. In addition, there had been five cases of premature withdrawal of gear boxes which were sent abroad for repair at a cost of Rs 3.44 lakhs. Besides, 28 aggregates amounting to Rs 10.72 lakhs were also withdrawn prematurely and sent abroad for repair. According to the Ministry, none of these items have been received back after repair.

In order to provide training to IAF personnel on second line maintenance activities of the aircraft, a team of specialists from the manufacturers was deputed to India under a contract concluded in July 1986. The team so deputed did not have the experts who could provide training on bay-servicing of aggregates in the laboratories which which was part of the contract. The deputation of another team of specialists from the manufacturers was, therefore, sanctioned in August 1988 to complete the residual work.

The entire expenditure on the second team should have been met by the manufacturers in view of the failure on their part. However, these had to be borne by the IAF due to ambiguities in the contract provisions. Air Headquarters stated in December 1988 that action would be taken to stipulate an appropriate clause in future contracts to avoid such situations.

The case reveals avoidable contracting delays resulting in an extra expenditure of Rs 2.94 crore in addition to a poor aircraft utilisation rate arising out of the absence of any preventive measures to avoid the high rate of premature withdrawals of engines and rotables. There would appear to be considerable scope for improved induction planning including a strong maintenance element.

IAF Jet Crash

Thirty-four persons were killed in eight defence aircraft crashes in the first three months of this year.

See also Accidents: Air.

INDIAN AIRLINES

Airbus A-320

The operation of the Airbus A-320 fleet of the Indian Airlines expected to be resumed in a phased manner.

Government had accepted the recommendations of the Ramdas Committee and Indian Airlines was taking the necessary steps and was fully geared to implement the recommendations of the committee

Bird Hits

The loss suffered by the Indian Airlines due to incidents of bird hits on aircraft rose from Rs 1.73 crore in 1988 to Rs 5.91 crore in 1989.

See also Accidents : Air

INDIAN COUNCIL OF AGRICULTURAL RESEARCH (ICAR)

The Comptroller and Auditor General of India has complained of delays in deciding on projects sponsored by ICAR, abandonment of projects and also non-provision of facilities after sanction has been accorded. A test check revealed 206 scientists in 69 projects were inadequately deployed and the total infructuous expenditure amounted to Rs 4.51 crore.

The Agro-Energy Centre was set up in November 1983 to conduct research on bio-mass production, utilisation, socio-economic studies and technology transfer with a sanctioned strength of 16 scientists, eight technical personnel, eight administrative and five supporting staff. Up to April 1988, the Centre had spent Rs 28.97 lakh inclusive of Rs 19.90 lakh on pay and allowances against the budget provision of Rs 50.88 lakh. Budget and Research Committee recommended (October 1984) five research projects and Research Council approved only one project in February 1986 though it had been started 16 months earlier without approval.

Sixty-eight scientists of 19 divisions/units were not engaged on any research project for periods ranging up to eight years either because they did not submit any research project or because the project submitted by them were not approved by the Research Council. This led to an infructuous expenditure of Rs 85.77 lakh.

Due to overall reorganisation, 12 scientists in six divisions had been working on projects not approved. The scientists had continued the work due to delay or non-receipt of any communication from the Research Council.

INDIAN COUNCIL OF PHILOSOPHICAL RESEARCH (ICPR)

The Comptroller and Auditor General of India has come down heavily ICPR pointing out several irregularities in its working.

Council's Chairman and Director (Planning and Research) visited Britain and the United States to purchase books worth Rs 40,000 incurring an expenditure of Rs 89,000 lakh in March, 1988.

The Governing body of the Council had in July, 1986 decided that foreign books would be purchased by placing orders with a reputed foreign bookseller.

In March 1988 the Council purchased UNESCO coupons worth Rs four lakh for purchasing books from Britain and the USA. Contrary to the decision of the governing body, the two officials went abroad to buy books worth Rs 40,000 lakh and then purchased books worth Rs 3,60 lakh in India after their return.

Government approval for the visit abroad was not obtained. Further, the two officials did not submit travailling allowance bills but only furnished utilisation certificates for $ 1320 each paid to them in foreign exchange.

The governing body of the council did not meet regularly and at prescribed intervals. Copies of the proceedings of the meetings were never sent to the Government as required.

The administrative expenditure of the council exceeded the ceiling in all the seven years from 1981-82 to 1988-89, except 1983-84.

In 68 cases, the chairman sanctioned Rs 14.44 lakh for extension of fellowships, seminars, travel grants and so on without the prior approval of the research project committee. Utilisation certificates for Rs 13.20 lakh paid for such purposes were not received.

There was no systematic reporting of the academic activities by the national fellows. Fellowships were awarded in 20 cases without them being affiliated to an approved university or institution.

The council did not publish the proceedings of any of the three major seminars held by it during 1981-88. Utilisation certificates for Rs 8.24 lakh paid to other organisations for conducting seminars during 1984-89 were not received.

INDIAN MISSIONS ABROAD

The Comptroller and Auditor General of India has raised objections against an expense of 3.2 million pounds by the Indian High Commission in London on renovating a building as a cultural centre.

The objections cover expenses during much of the period when Mr Maharaj Krishan Rasgotra was High Commissioner.

The money was spent without prior approval of the Government. Rules and procedures were violated in the selection of contractors. The contract went to a firm that was not on the approved list. This was a serious irregularity.

The renovation work on the building at 8, South Audley Street, is now complete. Plans are being finalised to set up a cultural centre on the ground floor and the first floor. The second and third floors are being used as residence for some High Commission officials.

The building in the fashionable Mayfair area is Indian Government property. It was used in the fifties and sixties for consular services of the high Commission. A decision to renovate the dilapidated building was reportedly taken when Mrs Indira Gandhi was Prime Minister.

The project has been dogged by controversy every now and then. At one stage officials proposed setting up the place as a library. But renovation work to convert into a cultural centre continued.

The building is now set to house a cultural centre to rival the Bharatiya

Vidya Bhawan in London. A director is expected to be appointed shortly.

But it is still not clear how some million pounds were spent without Government sanction on a contract awarded to an approved firm.

INDIAN NATIONAL SATELLITE (INSAT) SYSTEM

India's INSAT-ID communication satellite was launched successfully on June 12, 1990 from Cape Canaveral. INSAT-ID is a multipurpose operational satellite similar to INSAT-IB and INSAT-IC, which are already in orbit. INSAT-ID is intended to replace INSAT-IB, which is coming to the end of its useful life.

The satellite, built by Ford Aerospace Communication Corporation, was launched by one of McDonnel Douglas Corporation's three-stage Delta rockets, the reliable workhorse of the space programme.

INSAT system is a joint venture of the Department of Space, Department of Telecommunications, the Indian Meteorological Department, All India Radio and Doordarshan, INSAT is intended to provide operational services in telecommunications, meteorological observations and data relay, nationwide television broadcasting, radio and television programme distribution, disaster warning for coastal areas, and so on.

The INSAT satellites are maintained and controlled from the INSAT Master Control Facility (MCF) at Hassan in Karnataka.

INSAT-IA has an estimated life of about seven years.

The successful launch took place at 11.22 a.m. IST from the Eastern Space and Missiles Centre, Florida, USA. The Delta launch vehicle delivered the INSAT-ID spacecraft into a geosynchronous transfer orbit of 39.480.5 km 168.29 km with an inclination of 26.96 degrees.

The injection of INSAT-ID occurred over the continent of Africa at 11.46 a.m. IST. The Master Control Facility (MCF) at Hassan in Karnataka acquired the satellite telemetry signal at 12.03 p.m. IST. The separation of INSAT-ID from the payload assist module, which is the last stage of rocket, was commanded from Hassan at 12.38 p.m. IST. Thereafter, a check on the satellite's health was performed and the satellite was despun and commanded to the 'sun acquisition mode' whereby the west face of the satellite was oriented towards the sun.

The solar panels No. 4 and 5 were released at 1.03 p.m. IST. Since then, precise determination of orbital parameters and calibration of the Digital Integrating Rate Assembly (DIRA)—which provides attitude data during the apogee motor firings—are going on.

The period of revolution in the transfer orbit is about 11 hours. Towards the end of the first orbit, the satellite will go out of radio-visibility from Hassan. Using support from a global network of tracking stations, the Master Control Facility at Hassan will be in constant reception of all spacecraft telemetry at all times. The operations at MCF, Hassan, are being carried out by the joint team of scientists from the Department of Space and Ford Aerospace Corporation the prime contractor for the INSAT-1 spacecraft.

The signal reaquisition from Hassan was required occur in the third orbit. The apogee motor firings to raise the orbit perigee and orbit-trimmings are planned thereafter. The INSAT-ID satellite is slated to achieve, under nominal conditions, its position at 83 degrees East within 7-8 days. After this, further deployments of full solar array, C/S Antenna and the solar sail will be taken up. After successful completion of all deployments, on-orbit testing of the satellite will be performed on completion of which, the satellite will be declared ready for operational use some time around the first week of July.

INSAT-ID is the last of the first generation multipurpose satellite series, which has been fabricated by FAC of USA. INSAT-ID after fully operationalised, will provide communication, TV broadcasting, radio networking and meteorological services to the entire country.

It has like its predecessors, 12 C-Band transponders, 2 high-power C-band transponders and a Very High Resolution Radiometer (VHRR), which can take weather imageries both in the optical and in infrared bands. This launch has also given breathing time to the satellite communication network till the advent of the INSAT-2 series in a couple of years and reduce the dependence on leased capability from foreign countries. While INSAT-IB continues to be officially pronounced "alive", the Indian satellite network, according to reliable sources, is being run entirely on hired transponder capability aboard the Arabsat and the Soviet stationary satellite.

Till now, the INSAT-1 system provides television coverage through over 250 ground stations and 7,600 direct receiving sets along with regional TV sets in many States. On the telecommunication front, it operates more than 4,320 two-way voice-circuits in 101 routes. Also included in the system are over 100 Earth stations and several captive networks run by public utilities and Government departments like the ONGC, ITI, Nicnet and others. The satellites also send valuable meteorological data in the form of scan photographs over the Indian subcontinent, apart from maintaining 100 disaster warning systems along the cycloneprone eastern coast. More than anything else, the 12 transponders hired on the Arabsat at the cost of 800,000 dollars per transponder a year might now be given up after a certain transition period in order to cut the outflow of foreign exchange.

The $80 million INSAT satellite is to become operational in mid-July, replacing one that was deployed in 1983 by the crew of the shuttle Challenger.

India has paid Rs 1,340 million to the Ford Aerospace Corporation to build the satellite and for insurance coverage and Rs 620 million to the McDonnel Douglas Corporation of the US to launch it.

The Delta launch is also the first commercial launch of the McDonnel Douglas after the US privatised its launch services last year.

The satellite originally was scheduled to ride a Delta rocket into orbit last June, but was damaged in a launch pad accident a week before liftoff. A cable on a crane broke.

Ford Aerospace, which built the satellite, returned the spacecraft to Palo Alton California in June 1989. Workers there replaced an antenna and made other repairs costing $10 million, said Mr Fred Stang, Director of Business Development for the company.

The earthquake that struck northern California in October caused further damage—about a dozen cells in the satellite's solar panels had to be replaced. The cost for that was about $150,000, Mr Stang said.

Ford Aerospace has insured the satellite for any problems while it is on the ground. The Indian Government, which also has insurance, assumes responsibility after liftoff.

Because of the year's delay, India has had to lease the services of other satellite communications companies.

The new spacecraft is the fourth and final in a series of India's INSAT-1 satellites. An updated version, the INSAT 2, will be carried into space next year by Europe's Ariane rocket.

Barring to remote sensing satellite, IRS-IA, the other ISRO projects have not met with success during the past few years. The first two developmental flights of the Augmented Satellite Launch Vehicle (ASLY) in 1977 and 1988 failed. The INSAT-IC satellite suffered a major short circuit after its launch in 1988 and a freak accident put off the INSAT-ID launch by an year.

Ninth from Foreign Soil

INSAT-ID is the ninth Indian satellite to be launched from foreign soil.

The three-in-one satellite is the third to enter its space home from the United States.

While four satellites went up from the Soviet Union, two others were launched from the picturesque Kourou Islands in French Guyana.

Following are the satellites launched from abroad: Aryabhatta (1975-USSR), Bhaskara-I (1979-USSR), APPLE (1981-Kourou), Bhaskara-II (1981-USSR), INSAT-1A (1982-USA), INSAT-IB (1983-USA), Indian Remote-Sensing Satellite IRS-IA (1988-USSR) and INSAT-1C (1988 Kourou).

(Source: N.C. Menon)

INDIAN PEACE KEEPING FORCE (IPKF)

As many as 1157 persons (including 34 personnel reported missing and presumed killed) were killed in the (IPKF) in Sri Lanka.

INDIRA GANDHI NATIONAL CENTRE FOR ARTS (IGNCA)

IGNCA was set up as a fully autonomous trust in March 1987, following a Government decision of early 1985 for setting up a memorial institution named after Mrs Indira Gandhi for culture and arts. A separate department of arts was solely created for this purpose.

The IGNCA trust was initially constituted with seven trustees and later two more trustees were added to make the current trust composition of nine members, Mr Rajiv Gandhi is president of the trust. The other members are President R.Venkataraman, Mr P.V. Narasimha Rao, the Minister of State for Finance is an ex-officio member, Mrs Pupul Jayakar, Mr H.Y. Sharda Prasad, Mr Abid Hussain, Mr M.S. Subhalakshmi and Dr (Mrs) Kapila Vatsyayan as member secretary.

The Government provided corpus fund of Rs 25 crore to the fully autonomous trust of (IGNCA), another Rs 14 crore from Plan allocation for the Department of Arts, and hand-picked nine members of the trust.

The conceptual plan is stated to have been completed following the transfer to Rs 25 crore as corpus fund as well as of Rs 14 crore for the building project from the seventh Plan allocations. The trust has invested the Rs 25 crore corpus fund in long-term deposits to earn interest for financing its academic programmes.

Following instructions from the Cabinet secretariat, former director-general of Capart S.M. Patankar and an additional secretary of the Law Ministry inquired into the funding of IGNCA and its building project.

Dr Patankar, who submitted his report recently, has advised the Government to modify the scope of the trust project as well as its working and structure. This advice is based on Dr Patankar's assessment that the IGNCA trust deed has given an unusually long and secure tenure of 10 years to all the trustees once they are appointed by the Government. It has also empowered one-third of the trustees to nominate their successors on retirement after 10 years, the report has added.

Pointing out that there is no provision in the trust deed for removing its existing members, the Patankar report has highlighted the absolute absence of any institutional arrangement to facilitate Parliament to review the working of the trust which has been given Rs 25 crore from the tax-payers' money and promised a funding of Rs 100 to Rs 150 crore for constructing a group of buildings in 23 acres of prime land near the Central Vista in the Capital.

The report has, therefore, observed that a situation has been created whereby Parliament has lost control to oversee the utilisation of public funds. It has added that the present trust composition is partisan because it does not include any representative from other segments of society Dr patankar has finally concluded that the present composition enables the trust to exercise power in a manner that may be injurious to public interest.

The report has advised the Government to avail some of the safeguards contained in the trust deed to prevent any possible abuse of power or mismanagement. One of the safeguards is that the President of India as the Visitor of IGNCA may set up a review committee on the working of the organisation and the recommendations of the committee as accepted by the Government in consultation with the trust shall be binding on the trustees.

(Source: Mukul Kumar Shukla)

INDO-SOVIET TRADE

Indo-Soviet Trade Plan for 1990 envisaged turnover of Rs 8,800 crore, comprising Indian exports to USSR of Rs 5,300 crore and imports from the USSR into India of Rs 3500 crore.

INDUSTRIAL HOUSES

Total Assets, Profit before Tax and Total Income in 1988-89 (Accounting Year ending April 1988 to March, 1989) belonging to the 20 Largest Business Houses Ranked According to their Assets

Sl. No.	Industrial	Assets	Profit before tax	Total Income (including turnover in crores)
1.	Birla	6974.06	301.59	6852.33
2.	Tata	6621.38	439.18	6779.97
3.	Reliance	3241.24	84.95	1227.09
4.	J.K. Singhania	1828.75	36.72	1587.12
5.	Thapar	1762.52	49.70	1493.43
6.	Mafatlal	1296.55	88.49	1687.86
7.	Bajaj	1228.37	60.61	1196.59
8.	Modi	1192.34	15.56	1828.27
9.	Larsen & Toubro	1130.33	56.94	1214.31
10.	M.A. Chidambaram	1032.23	29.19	747.26
11.	T.V.S. Iyengar	020.06	58.57	1080.90
12.	Hindustan Lever	924.85	180.21	2510.08
13.	A.C.C.	909.13	–23.96	730.01
14.	Shri Ram	799.17	–2.78	1128.72
15.	I.T.C.	742.19	66.58	1796.95
16.	United Breweries	715.71	29.46	726.11
17.	I.C.I.	674.46	54.17	1239.03
18,	Bangur	657.41	-15.11	771.96
19.	Kirloskar	633.07	25.87	719.26.
20.	Walchand	625.75	20.33	686.20

Note:

* Includes Southern Petrochemical Industries Corpn. Ltd. which was included in M.A. Chidambaram group in 1985.

INDUSTRIAL POLICY

The paper on "Policy measures for the promotion of small-scale and agro-based industries and changes in procedures for industrial approvals" tabled in the Lok Sabha on May 31, 1990, was under the consideration of the Government. The production performance of various infrastructure industries during April-November, 1990, was below target.

INDUSTRIAL PRODUCTION

Associated Chambers of Commerce NS Industry of India (ASSOCHAM) survey of 'Prospects for Industry in 1990-91', which covers selected industries with combine weightage of about 30 per cent in the index of industrial production and over 40 per cent in the manufacturing output, predicts significant improvement in the availability of steel, cement, jute, paper, textiles, fertilisers, etc. IN 1990-91 over.

Steel, an essential input in the engineering industry, has been registering low production. In 1988-89, steel plants failed to achieve the targeted production 9.3 million tonnes of crude steel and 7.8 million tonnes of saleable steel. The actual production of crude steel was only 8.27 million tonnes (89 per cent of the target) and saleable steel 7.0 million tonnes (90 per cent of the target). The shortfall was mainly on account of inadequate supply of coal and power and inconsistent quality of coal.

As a result of measures recently announced by the Government, steel production in smaller and medium private sector units is likely to increase resulting in significant reduction in imports of scrap and pig iron in 1990-91.

The new steel policy also opens the blast furnace technology to the private sector subject to the ceiling of 2.5 lakh tonnes capacity per annum. The policy also seeks to develop the sponge iron industry. Production of 5 million tonnes of sponge iron, a substitute to scrap, has been planned by the end of the 8th Plan. This would reduce dependence on imports of scrap.

Cement production in 1989-90 was 44.71 million tonnes. It fell short of the 7th Plan target of 49 million tonnes. The shortfall was mainly because of lower off-take by the Government departments and shortages of coal, wagons, power, etc.

With the Government's stress on housing, canal lining, road construction, etc., the cement industry is expected to receive fresh impetus. By the end of the 8th Plan, the demand of cement may vary between 66 and 69 million tonnes. Production is expected to grow at the annual rate of 8 to 9 per cent.

Production of the jute textiles in

1989-90 fell by four per cent from 1147.4 to 1125.2 thousand tonnes. A dozen jute mills had to close down because of shortage of jute. However, production is expected to improve significantly because of the year's bumper crop. Stricter implementation of mandatory packaging orders and the recently introduced full fibre flexibility scheme will help to rejuvenate the industry.

In 1990-91, production of paper and paper board is expected around 90.08 lakh tonnes as against 18.2 lakh tonnes in 1989-90. The expansion in agro-based mills holds greater promise to meet the projected demand of 32 lakh tonnes by the end of the century. This is because of the abundant availability of the agricultural residues. As against this, wood based paper units are faced with inadequate raw material supplies on account of depleting forest resources.

Cotton is the chief raw material for the textile industry. Though the current year's crop has been estimated to reach a new high of 127.5 lakh bales as against 106 lakh bales last year, the decision to raise the minimum export quota of raw cotton will restrict domestic availability. Caprolactum used for the production of nylon yarn is also in short supply. If the textile industry is modernised and basic raw materials are supplied, production as well as exports could very substantially increase. The current sugar season as a whole is likely to see a production of 105 lakh tonnes, 20 per cent higher than the production last year.

Presently only about one per cent of the total fruit and vegetable production in the country is processed. In 1990-91, the Ministry of Food Processing plans to adopt various schemes to provide further fillip to the industry by strengthening linkages between growers and processors, emphasising quality control and developing infrastructural facilities.

For the current year production target for fertilisers has been fixed 11.6 per cent higher than last year's 9.7 million tonnes, of which 7.1 million tonnes is nitrogenous fertilisers and 2.6 million tonnes phosphatic fertilisers, Growth may be dampened considering the effect of the recently announced curbs on consumption of petroleum and petrochemical products.

The target of 8 per cent industrial growth in 1990-91 is much below the capability of the industry. With the capacity utilisation of around 77 per cent unsatiated internal demand, development of export markets, cutting down costs and improvement in quality, the country is capable of achieving 10 to 11 per cent industrial growth in industry on a sustained basis.

The new industrial policy could be a stepping stone for growth particularly for the small and medium sized units. Growth could be even well beyond 10 to 11 per cent, if shackles on capacity and diversification on MRTP houses and FERA companies are removed.

(Source: *Hindustan Times*)

INDUSTRIAL PRODUCTION INDEX

According to the quick estimate of Index of Industrial Production (IIP) with base 1980-81, issued by the Central Statistical Organisation, the general index for the month of January, 1991, stands at 225.3, which is higher by four per cent as compared to the corresponding month of 1990.

Taking into account the IIP figures released up to December, 1990 and the quick estimate of IIP for the month of January 1991, the average Index for the first ten months of the financial year 1990-91 works out to be 205, which is higher by 8.4 per cent as compared to the index of 189.1 for the corresponding period of the previous year.

INDUSTRIAL RELATIONS

Arbitration and Legislation

A 10-member committee was constituted by the Indian Labour Conference with the directive to "formulate specific proposals for further legislation to amend the Industrial Dispute Act, 1947".

The Bipartite Committee on new industrial relations law, headed by G.Ramanujam unanimously agreed that the main thrust of the new legislation should be on bipartism through direct negotiating council.

While bipartite negotiations will occupy the pride of place in the new industrial relations law, the Committee provided for voluntary arbitration wherever dipartite negotiations failed.

The voluntary arbitration either by a single arbitrator or by a board of arbitrators should not have the trappings of the present day Industrial Tribunals.

The arbitrators award should not be challenged in any court except where the arbitrator had acted without jurisdiction or on account of errors apparent on the face of the record or where the arbitrator is guilty of misconduct. The arbitration and promotion board should be revived so that voluntary arbitration becomes a normal way of resolving industrial disputes that could not be settled through bipartite negotiations.

Wherever direct negotiations fail and voluntary arbitration is not available for any reason, the act will provide for conciliation, high powered mediation and adjudication by the Industrial Relations Commission but wanted the existing machinery to continue with some changes. Strikes should be a rare occurrence and lockout even rarer. These should be used as a last resort. The interest of the society at large should be kept in view before the parties resort to either of these methods.

The minimum strength for registration of trade union should be 10 per cent of the employees in an establishment or 100 whichever is less, subject to a minimum of seven members. The right to form trade unions should be available to all employees regardless of their status, salary, wage or the character of the employees. Trade union rights should also continue to be available to the employers.

In the organised sector, registration of trade unions should be made compulsory. The National Federations and National Centres are exempted from such compulsory registrations. Election of office bearers and Committee members shall be once in three years or earlier if the constitution provides for it. Elections may be held either by show of hands or by secret ballot as the members may decide. While labour members desired that dual membership should be discouraged, employer members wanted it to be prohibited.

No trade union shall be registered if its membership is restricted to a particular craft, occupation, caste, creed, community, race, religion, or persons originating from a particular region. Registration of trade union should be made within 45 days of receipts of application. In case of refusal to register, an appeal will lie to the labour court at the Centre or States as the case may be.

All intra-union disputes shall be settled by the trade union centre to which the union is affiliated.

So far as the outsiders are concerned, it is the right of the worker to choose the leader he wants. Even so, not more than one third can be outsiders as office bearers and members of the executive committee put together.

Ministers at the Centre or in the States shall be ineligible to hold any office of trade union.

The title of the law should be changed from the Industrial Disputes Act to Industrial Relations Act. It will provide appropriate machinery to settle disputes by direct bipartite negotiations, voluntary arbitration, conciliation, mediation and adjudication, strikes and lockouts.

All employed persons regardless of the character of the employer or the destination of profits should be covered by the new law.

The Labour side however, wanted that definition of the term "industry" should be defined in the Industrial Disputes (Amendment) Act, 1982. The employers side, however, wanted the 1982 definition to remain unaltered. Regarding the definition of the term "employer", while the labour side wanted that the Act should cover all employees, the employers side desired that the definition of the term workmen as contained in the Industrial Disputes Act, 1982 should remain.

All employer-members and some labour-members supported the setting up of an Industrial Relations Commission. In view of the great importance assigned to Dipartism, no unilateral action can be undertaken by either the management or employees.

The Labour side had no objection for

retraining of workers for effective utilisation of services and wanted a higher retrenchment compensation.

The employer must share with his employees in the Negotiating Council all information about the financial aspect of the establishment even during normal times. The compensation paid to the workers should take precedence over all other claims on the employer.

(Source: *Hindustan Times*)

Disputes

37,974 industrial disputes and applications were pending before the labour courts and industrial tribunals in Delhi as on June 30, 1990.

On an average about 6,000 industrial disputes and applications were disposed of in a year in Delhi.

Strikes

The industrial relations situation in the country remained stable during 1989. The number of strikes and lockouts declined from 1156 to 1110 and the number of mandays lost from 23.10 million to 16.37 million.

Workmen's Compensation and Benefits

The working group on industrial relations has recommended setting up of six new Central Industrial Tribunals during the Eighth Plan period. According to the Annual Report of the Ministry of Labour, greater stress was laid on the social security and welfare. The Workmen's Compensation Act provides compensation to workmen in case of disablement, temporary as well as permanent and their dependents in the case of death. The minimum amount of compensation for death is Rs 20,000 and for permanent disablement Rs 24,000.

The maximum amount of compensation for permanent disablement can go up to Rs 1,14,000 while that for death can go up to Rs 90.000 depending on the wages and age of the workman at the time of his death. In case of temporary disablement, compensation at the rate of 50 per cent of wages is payable for a maximum period of five years.

The Gratuity Act provides for the payment of gratuity on termination of employment or superannuation, retirement or resignation or death or disablement subject to completion of continuous service of not less than five years. The condition of continuous service of five years does not, however, apply in the case of death or disablement. The Act provides for the payment of gratuity at the rate of 15 days for every completed year of service or part thereof in excess of six months. The maximum ceiling for payment of gratuity is Rs 50,000.

Similarly, benefits are provided to workmen under the Employees' Family Pension Scheme. The benefits extended out of the scheme are family pension, life insurance and retirement-cum-withdrawal benefits. The rate of family pension ranges between Rs 225 and Rs 750 depending on the pay of the member at the time of his death. In addition, a lumpsum of payment of Rs 5,000 is made to the first recipient of the family pension. The surviving members are entitled to retirement-cum-withdrawal benefits ranging between Rs 110 and Rs 400 (for one year membership) and Rs 9000 to Rs 19,825 (for 40 years membership) depending on the pay of the member at the time of his death.

(Source: *Hindustan Times*)

Labour Cases in Delhi

As many as 42,216 cases are pending in labour courts in Delhi, of which 3,911 are pending for over five years.

Two more labour courts were sanctioned for Delhi in 1990-91, which will start functioning when judicial officers are made available by the Delhi High Court.

INDUSTRIAL SECTOR

During the financial year 1990-91, figures available up to December 1990 show that the industrial sector recorded a growth of 8.9 per cent, over the same period in 1989-90.

In the manufacturing sector, the industry groups which have shown a good performance in the Seventh Plan period include electrical machinery, chemical and chemical products, leather and fur products, paper and paper products and basic metals. However, groups like beverages, tobacco products, jute and wood and wood products showed a decelerated performance.

During April-December 1990, mining and quarrying recorded a growth rate of 1.7 per cent while power generation showed a growth rate of about 7.1 per cent. In the manufacturing sector, the growth rate has increased to 10.4 per cent in April-December, 1990 compared to 4.4 per cent over the same period during the previous year. Some of the industry groups which contributed to this increase include food products, basic metals, electrical machinery, wood and wood products, paper and paper products and cotton textiles.

Approvals for 1990-91 at Rs 11,863 crore exceeded the capital approvals for the previous year (1989-90) by 46.1 per cent. Approvals in 1990-91 (April-December) at Rs 9,564.1 crore showed a decline of 13.8 per cent compared to the same period in 1989-90. The financial assistance sanctioned by all financial institutions for the year 1990-91 (April-December) was Rs 12,396 crore which showed a rise of 22.8 per cent over the corresponding period in 1989-90.

Disbursements by all India financial institutions, which stood at Rs 7,031.1 crore during 1990-91 (April-December), registered an increase of 24.3 per cent over the same period in the previous year.

The number of small scale industrial units are estimated to have gone up to 418.26 lakh at the end of 1989-90 showing an increase of about 6.7 per cent. Over the same period, employment in this sector is estimated to have increased from 113 lakhs persons to about 119.60 lakh persons making an increase of 5.8 per cent. Production in the small scale sector has been estimated at Rs 1,32,320 crore during 1989-90 compared to Rs 1,06,400 crore during 1989-90 at current prices.

(Source: *Hindustan Times*)

See also Production Index.

INDUSTRIAL UNITS/UNDERTAKINGS

Delhi Units

There are about 81,000 industrial units in Delhi, employing approximately 7.29 lakh workers. As December 31, 1990 there were 5,647 industrial units registered under the Factories Act, all of which were complying with the labour laws.

Takeover

The management of 55 industrial undertakings has been taken over by the Government so far.

Experience gained from managing the taken over units had not been very satisfactory.

INFANT MORTALITY

Infant deaths in India is estimated at 21.46 lakh every year or about 5,900 per day.

In West Bengal, infant deaths number 1.15 lakh annually or 315 per day.

The infant mortality rate as per the sample registration system of the registrar general of India was 94 per thousand live births. For West Bengal it was 70 per thousand live berths.

Children born today in an environment of poverty in Delhi are many more times likely to die before the age of five years and suffer more illness due to infections and malnutrition than those born in better-off families, according to a UNICEF report entitled *'The Invisible Child'*.

Delhi had one of the highest per capita incomes in the country (Rs 5,464 at 1985-86 prices) and an infant mortality rate (IMR) of 39.69 (1985). Yet, 70 per cent of household incomes were less than Rs 500 per month in the slums surveyed, and infant mortality rates varied greatly. One study found the average IMR to be 86, while another study found it to be 112, with IMRs as high as 200 in some slum areas.

(Source: *Hindustan Times*)

INFLATION

The rate of inflation on the basis of the average wholesale price index for the first quarter of 1990 works more at 8.2 per cent in 1990.

The inflation rate touched a double digit figure compared to 7.6 per cent in November last year.

See also Price Rise.

INFORMATION TECHNOLOGY INDUSTRY

A decision to prepare the plan was taken at a meeting of the executive council of Manufactures Associations for Information Technology (MAIT), the apex body of over 100 leading Indian manufacturers of computer hardware, software and peripherals.

The computer industry has recorded a growth of 45.54 per cent during 1989-90, from a turnover of Rs 1070 crore in 1988-89 to Rs 1,558 crore in 1989-90. Of the total, the turnover of hardware industry amounted to Rs 784,70 crore, recording a growth of 42.24 per cent from Rs 526.35 crore in the previous year.

Software has recorded a growth of 88,72 per cent to Rs 97.51 crore in 1989-90 from Rs 51.67 crore in 1988-89.

Another vital segment of the industry, i.e., training, recorded a growth of 151.50 per cent, with a turnover of Rs 67.66 crore in 1989-90 from Rs 26.87 crore in 1988-89.

MAIT had set up a computer training division, which would grant recognition to institutes on the basis of standards in curriculum and equipment set by it.

India's exports in the field of computers during 1989-90 amounted to Rs 226.75 crore, recording a growth of 88.25 per cent over the previous year Of this exports of hardware a bulk of it to the Soviet Union, amounted to Rs 66.20 crore.

MAIT is seeking to promote exports by creating an image of India by taking delegations abroad and participating en exhibitions. MAIT would also make use of the offices abroad of Federation of Indian Export Organizations, of which it is an Associate.

(Source: *PTI Release*)

INTEGRATED CHILD DEVELOPMENT SCHEME (ICDS)

There are 117.82 lakh children and 22.71 lakh mothers receiving supplementary nutrition and 64.36 lakh children attending pre-school education under ICDS which completes 15 years of its implementation on October 2.

During these years a total of Rs 800.39 crore has been spent by way of raising the nutrition status, enhancing the awareness of mothers and improving the immunisation coverage of children and mothers in the rural, tribal, drought and flood-prone areas.

The Department of Women and Children Development has suggested a number of programmes relating to organisation of Bal Shivirs and melas at block and circle-level during the week, essay-competition among anganwadi workers, helpers and programme functionaries.

(Source: *Hindustan Times*)

The Comptroller and Auditor General has noticed several deficiencies in the implementation of ICDS. During the Seventh Five Year Plan period (1985-90), an expenditure of Rs 660.28 crore was incurred outstripping the Plan outlay of Rs 520 crore. In sharp contrast, the physical progress of 2,236 projects by March 1990 was way behind the Plan target of 3,019 projects. Out of the 2,236 projects sanctioned, only 1,617 projects were operational and of these 215 projects were interrupted.

The focal point of ICDS services is an "anganwadi". Anganwadis had not been set up in accordance with the norms. Most of the projects and anganwadis, in some of the State, had no safe and potable water supply, which was essential for improvement of health status of the target group.

The implementation of the scheme was adversely affected on account of delays in supply and or non-supply of the essential equipment like first-aid boxes, medicine kits, weighing scales etc. to the projects. Consequently the benefits failed to reach the target group.

Under supplementary nutrition, shortfalls upto 80 per cent occurred in coverage of women and children and in some cases supplementary nutrition was not at all provided. Feeding suffered from frequent interruptions, for periods exceeding two years in some cases, and was provided for 28 days only against the norms of 300 days in some of the States test-checked.

The quality of supplementary nutrition was below the prescribed norms: therapeutic food, rich in calories and proteins, required to be supplied to severely malnourished children, had not been supplied in several States. In nine States, cases of supply of substandard and infected food were noticed.

Although the necessity for development of a system of cheap and hygienically prepared ready-to-eat food had been recognised, no action in the matter had been taken.

As a result of inadequate, interrupted and sub-standard quality of supplementary nutrition, the desired effect of raising the health status of beneficiaries could not be achieved.

INTEGRATED RURAL DEVELOPMENT PROGRAMME (IRDP)

More than 34 million families have been assisted between April 1980 and February, 1990 under the IRDP.

This also includes over five million families of the sixth plan period assisted for supplementary dose of assistance during the Seventh Plan period.

INTERNATIONAL LOAN MARKET

India is faced with a major crisis in its commercial borrowings programme in the international market following the latest assessment made by the international agencies that the Indian economy is on a downhill course and the present credit rating is no more valid.

Immediate impact of this approach has been felt in the case of the leading public sector bodies including the financial institutions who have been scouting in recent weeks in the international market for syndicated loans. The Oil and Natural Gas Commission (ONGC), the cash rich creditworthy public sector unit, has failed to get its $150 million loan syndicated by the Japanese banks, who till the other day were wooing the Indian parties for such loans.

Moody's Investor Services, a leading US rating agency, has placed India on 'watch' and this means the stage before lowering the rating in the international financial market. This has been for the first time that the Indian financial institutions have been put on watch.

The international bankers are losing confidence in the capacity of the present Government to manage effectively its balance of payments situation. The present attitude of the international bankers stands in contrast to the observations made by the World Bank in its country report for India presented to the meeting of the Aid India Consortium .

The World Bank report says that publicly guaranteed medium and long term debt would rise from about US$ 56.8 billion at the end of 1989-90 to about $89.1 billion at the end of 1994-95, while total external debt would increase from US $ 64 billion to $ 100.6 billion. The percentage of total debt on concessional terms would decline from about 40 per cent to about 34 per cent. The debt export ratio would decline significantly from 2.6 in 1989-90 to 2.3 in 1994-95 and the debt service ratio sustained by a bulge in scheduled payments of loans from private creditors beginning in 1992-93, would decline more slowly from 28.1 per cent to 27.3 per cent. After this bulge, which is related to high levels of commercial borrowing of 1988-89 and 1989-90 is digested the debt service ratio would be expected to decline more rapidly in the late '90s.

Thus the World Bank report predicts a decline in debt service ratio in the early years of the Eighth Plan period but the international bankers have been more influenced by the short term factors in the Indian economy in assessing India's credit rating.

(Source: Nitya Chakravorty)

INVESTMENT TRENDS

The investment climate in the country for the first nine months of 1989-90 has shown a mixed trend according to a study undertaken by the Federation of Indian Chambers of Commerce and Industry (FICCI) on the investment environment in the country.

While indicators of investment, such as, capital issues, import of capital goods, cases approved under the schemes of Minimum Economic Scales

(MES), etc., have shown a positive trends indicating an encouraging investment climate, the sanctions and disbursement by the major financial institutions, letters of intent issued, foreign collaborations approvals, etc., have shown a declining trend during the first nine months of 1989-90 as compared to the same period in the previous year.

The registrations issued by SIA under the scheme of delicensing has increased by 8 per cent and the cases approved under the scheme of MES scheme increased by 71 per cent. Significant increases have also been there in the capital issues approvals (100 per cent) capital goods import approval (29 per cent), cases approved under the scheme of broad banding (32 per cent), carry on business licence issue (36 per cent) and industrial licences by way of conversion of letter of intent/direct licences (5.5 per cent).

These positive factors, have contributed to the impressive growth rate registered by industry. In 1989-90 the industrial sector recorded a growth of 8.3 per cent and is expected to record a 9 per cent growth rate during 1990-91.

However, the aggregate financial assistance sanctioned and disbursed by the financial institutions in the first 9 months of 1989-90 has continued to decline as compared to the same period in the previous year and the trend is still persisting. According to the latest data available, the aggregate sanction amounted to Rs 1,600 crore during April-June 1990 compared to Rs 2,600 crore during the first quarter of 1989-90, registering a fall of nearly 40 per cent.

The aggregate disbursement declined to Rs 1,060 crore during the first quarter of the current year from Rs 1,400 crore during the corresponding period last year, registering a fall of 24 per cent. This is a disturbing trend since this has happened for the first time in several years. In fact, institutional sanctions have been growing at an average growth of around 20 per cent during the 1980. It is also disturbing that these institutions have any investment proposal of significant size under consideration. It is, therefore, expected that the declining trend in sanctions and disbursement may continue in the second quarter also, i.e., during July-September 1990.

The important reasons for the fall in sections and disbursement are tightening of scrutiny of projects for sanction of loans by the financial institutions and delay by other government clearing bodies in granting clearance, as also the increase dependency of the private sector on the capital market. This has shown by the fact that capital floatation by the private corporate sector totalled Rs 5,616.3 crore during the latest 12 months period ending June 1990. These issues were 41 per cent higher compared to such issues of Rs 3974 crore in 1988-89 and were more than three and half times the issues of Rs 1,589 crore during the year 1984-85.

Another indicator of investment is the production of capital goods. Imports have been showing an increase trend. Import of machinery and transport equipment increased from Rs 6,746 crore in 1988-89 to Rs 8,500 crore in 1989-90, recording an increase of 26 per cent. Similarly, the production of capital goods during 1989-90 registered a growth of 25 per cent compared to the growth of 9 per cent in 1988-89.

(Source: *Hindustan Times*)

Consortium Aid to India

(In $ Billion)

Year	Amount
1986-87	4.4
1987-88	5.4
1988-89	6.3
1989-90	6.7
1990-91	6.5

Investment in Public Sector Enterprises (PSE)

The total investment in the Central Public Sector Enterprises was Rs 85,563.78 crore as on March 31, 1989.

Out of the 222 public sector enterprises earned an estimated profit of Rs 2175.73 crore in the first nine months period of 1989-90, ending December 31, 1989.

Investment in Public Sector Enterprises: Non-Petroleum Group

(Rs in crores)

	1986-87	1987-88	1988-89
Investment	56498	64662	77181
Profit	—	—	417
Loss	–317	–142	—

IRON ORE

The sale value of Bailadila iron ore from Bastar region exported by the MMTC to Japan in 1989-90 was Rs 203.83 crore. During 1988-89 and 1987-88 it was Rs 163.42 crore and Rs 119.69 crore, respectively.

IRRIGATION

India with 56 million hectares topped the list for irrigation, followed by China with 47 million hectares, USA with 27 million hectares and the Soviet Union with 21 million hectares.

While global statistics of irrigation expansion are poor, it is estimated that the present rate of development of new irrigated areas is only about two million hectares a year or about half the levels reached in the period from 1950-1980.

Thought India has the largest irrigation system in the world but it is inefficient. The gap between creation of facilities and utilization is widening. At the end of the Seventh Plan there was more unutilized potential than was added during the Plan, according to an expert.

Even the quality and impact of the irrigation is far from satisfactory. Time and cost overruns are a poor reflection of irrigation planning

The nation had spent around Rs 30,000 crores on an irrigation potential of nearly 30 million additional hectares since 1947. However, drought and floods were not effectively controlled and combated. Due to incorrect project, planning and methods of storing water, there was even drinking water shortage in Goa, Kerala and Cherapunji.

Civil engineering approach to irrigation and water use projects had to be replaced by ecological and more holistic approach.

(Source: *Statesman)*

J

JAMA MASJID, DELHI

Conservation works of Jama Masjid, Delhi, have been taken up as a special case, considering its archaeological, architectural and historical importance. The Archaeological Survey of India spent Rs. 32,118 for its conservation in 1989-90.

JAWAHAR ROZGAR YOJNA (JRY)

The Comptroller and Auditor General of India has pulled up the Agriculture Ministry for the manner in which the ambitious Jawahar Rozgar Yojana was implemented without adequate fund and for failing to adhere to the prescribed criteria for release of Central assistance to States under the programme.

The total resources of Rs. 2,623.08 crore provided for the programme in 1989-90 were inadequate to achieve its objectives and could sustain employment for hardly 21 days on an average to one member of each rural family living below the poverty line.

Central assistance to six States and six Union Territories was released in excess of their entitlement while in the case of 19 States and one Union Territory, it was less.

Though geographical and other factors were cited as consideration for increased allocation, no fresh guidelines on the basis had been laid down.

The programme was launched from April, 1989, by merging the National Rural Employment Programme (NREP) and the Rural Landless Employment Guarantee Programme (RLEGP).

The main objective of the programme was to generate additional employment for the unemployed and under-employed in rural areas, with expenditure to be shared between the Centre and States on an 80:20 basis.

Though the overall utilisation was hardly 24 per cent of available funds, the second instalment of Central Assistance worth Rs. 1,112.60 crore was released prematurely to the districts in September 1989.

The prescribed conditions of 50 per cent utilisation of the available resources had been satisfied in only two States and one Union Territory.

The premature release of the second instalment only contributed to increased accumulation of unutilised balances with them.

Further, the unutilised resources of the NREP and the RLEGP totalling Rs. 422.48 crore available with the States and Union Territories were not reckoned while fixing employment generation targets for 1989-90.

Due to the launching of the new programme without adequate preparations, employment generated in the first four months was less as compared to the average in the corresponding period of the previous three years under NREP and RLEGP.

This was despite the availability of more resources for the programme.

Targets set for generation of employment were not achieved, employment generation was nil during the first quarter of 1989 in three of the four districts test-checked by audit in Uttar Pradesh.

In Karnataka, achievement was only 36 per cent in the districts test-checked while in Andhra Pradesh, only about 11 per cent of the target was achieved.

Identification of the targeted beneficiary families, essential for proper implementation, was not done in Andhra Pradesh, Bihar and Uttar Pradesh while in Karnataka and Rajasthan work was in progress.

As many as 20 of the States and Union Territories have failed to provide even 50 per cent of the targetted employment under the Jawahar Rozgar Yojana during 1989-90.

A provisional assessment of the performance of the scheme showed that Tamil Nadu was the lone State which succeeded in generating more mandays under the programme than the target with an achievement of 100.84 per cent.

As against this, Assam and Meghalaya among the States and Dadra and Nagar Haveli and Chandigarh among Union Territories turned out a poor performance. While Assam and Meghalaya could achieve only 19.23 and 19.32 per cent of the targets. Dadra and Nagar Haveli and Chandigarh could fulfil 14.32 and 15.96 per cent of the task, respectively.

(Source: *Hindustan Times*)

JOBS IN GULF

Some groups/companies dupe the people in the name of providing jobs in the Gulf countries.

FIRs have been lodged against 27 groups/persons involving the complaint of cheating of 70 persons during the last two years.

Prosecution has been sanctioned under the Emigration Act 1983 against 26 persons/groups. He said the Emigration Act 1983 and the rules framed thereof were designed to protect the emigrant workers against exploitation.

JOURNALISTS

An Interdepartmental Empowered Committee has been constituted by the Government to process the recommendations of the Ishwari Prasad Committee concerning working journalists.

JUDICIARY

The Government has accepted the policy recommendations of the Law Commission according to which one third of the judges in each High Court should be from any other State. Government had started appointing Chief Justices of High Courts from outside and some initial appointments of puisne judges from outside had also been made after the policy recommendations in the 80th report of the Commission.

As on April 30, 1990, against the actual strength of 25 judges, there were few Scheduled Castes judges in the Supreme Court.

On the same date, against the actual strength of 377 judges/additional judges, there were nine judges belonging to SCs and three judges to the STs in various High Courts.

There was one woman judge in the Supreme Court and 11 women judges out of 377 judges in various High Courts.

Vacancies

There were 50 vacancies in a total sanctioned strength of 468 judges in various High Courts and one vacancy in the Supreme Court as on 18.7.90. As on 1.8.90, there were only 27 vacancies left in the total sanctioned strength of 470 judges in the High Courts.

K

KASHMIR SITUATION

The Government clamped dusk-to-dawn curfew in the 5 km belt along the border and line of control besides stepping up vigil following reports that nearly 6,000 trained Kashmiri militants were amassed along the border to be pushed into Jammu and Kashmir. The Disturbed Areas Act and the Armed Forces (Special Powers) Act have been enforced in the border districts. In the first half of 1991, as many as 200 terrorists were apprehended and 133 killed on the border.

During the same period 94 security force personnel lost their lives in encounters and terrorist attacks.

During the last three months, eight tourists, seven Israelis and one Dutch, were abducted in the Valley by terrorists. No Indian tourist was reported to have been abducted.

Till June this year, more than 54,000 migrant families from the Valley, mostly non-Muslims, had been registered at Jammu and 18,000 at Delhi. Some families have taken shelter in UP, Haryana, Maharashtra, Gujarat, Himachal and Madhya Pradesh.

The Centre has asked the J&K Government to formulate a plan for the return of migrants to the Kashmir Valley in stages.

The State Government has been advised to make full security arrangements at the Hazratbal mosque in view of reports that Pakistan might try to provoke mass uprising in Kashmir with the clandestine removal of the hair of Prophet Mohammed from the mosque.

Over 72,000 Kashmiri families had been registered as migrants in Jammu and Delhi till June 1991.

Registration of Kashmiri migrants at places outside the Valley was started in the year 1990 only.

In Jammu, the migrants are being given cash assistance of Rs. 10 per day per head subject to a maximum of Rs. 1,000 per month per family of 4 or more persons, in addition to free ration of 9 kg rice, 2 kg atta and 1 kg sugar per head per month and accommodation in Government buildings/tents subject to availability. Free medical aid, blankets etc. have also been provided.

In Delhi, cash relief of Rs. 200 per person per month subject to a maximum of Rs. 800 per month per family of 4 or more persons is being provided to the migrants living outside the camps. Those living in camps, set up by Delhi Administration, are being paid cash relief of Rs. 125 per head per month subject to a maximum of Rs. 500 per month per family of 4 or more members in addition to one time kitchen and bedding kit and monthly dry ration costing about Rs. 500. Arrangements for regular visits of medical teams to the camps have also been made.

Instruction have been issued for admission of children to schools, colleges and professional institutions, transfer of bank and postal saving account, payment of leave salary, pension and also expeditious disposal of LIC claims etc.

According to the Government of Jammu and Kashmir, during 1990-91 an amount of Rs. 35.65 crore and during 1991-92 till date an amount of Rs. 18 crore has been spent on payment as cash assistance. Similarly, amount spent on distribution of free ration during 1990-91 was Rs. 8.60 crore, whereas during 1991-92 to date Rs. 6 crore have been spant. The Delhi Administration have been spent Rs. 3,79,50,873 on payment of cash relief to migrants during 1990 and 1991 (up to 30th June).

KILLINGS

See Punjab Killings; United Libration Front of Assam: ULFA Killings.

KNITWEAR

Export

Knitwears made in India are popular abroad, specially in the Soviet Union. They are being exported for the last two decades. Export exchange earnings from knitwears during 1989-90 was Rs. 782 crores. This is Rs. 291 crores more than in the previous year.

(Source: *Yojana*, July 1-15, 1990)

KUTIR JYOTI

Rs. 25 crore was provided by the Central Government in 1988-89 and 1989-90 for the "Kutir Jyoti" scheme, for providing a single-point light connection to the rural poor below the poverty line.

The service charges realised by the Rural Electrification Corporation which administered the "Kutir Jyoti" scheme, amount to Rs. 59 lakh, at the rate of Rs. 5 per connection.

KUWAIT

About 25,000 Indian nationals have returned to Kuwait after the end of the Gulf war. Indian nationals who left Iraq have not by thin returned. These were the only two Gulf States from which Indian nationals had to be evacuated.

L

LAND

Surplus Land

The total land declared surplus as on June 30, 1990 was 28.90 lakh hectares of which 24.85 lakh hectares had been taken possession and 18.59 lakh hectares distributed. Of the area not distributed so far, 5.05 lakh hectares were locked in litigation and 4.06 lakh hectares either reserved for public purpose or considered unfit for distribution. The net area available for distribution was 1.20 lakh hectares.

Land-holdings

The percentage of operational land holdings of small and marginal farmers of the total number of operational holding in the country was 18.4 and 57.8 respectively, according to 1985-86 Agricultural Census.

The net cultivated area of small and marginal farmers was 242 and 206.3 lakh hectares respectively.

Railway Land

35.67 hectares of railway land was under illegal possession in hilly areas.

About 635 hectares of railway land in hilly areas is under licence. Railway land is licensed for only those purposes connected with railway working, such as stacking of goods transported by rail, staff welfare measures. Licences granted prior to formation of this policy in 1984, even if they are not connected with railway working, are however, renewed annually till such time the land is required for railways development works.

Land Acquisition in Delhi

About 3500 bigha of land was acquired in the Union Territory during 1988 to 1990 and a sum of Rs 6.91 crore was paid as compensation.

LANGUAGES

Allocation of Funds

A sum of Rs 1859.80 lakh has been tentatively allocated for development of modern Indian languages including tribal languages for the current financial year.

LANKAN REFUGEES

See Sri Lankan Refugees

LEATHER

According to a study undertaken by PHD Chamber of Commere of Commerce and Industry CI on "India's exports to the East European Countries (EEC) of leather and leather products", India's export of leather and leather products stood at ECU 285 million in 1989 compared to ECU 282 million in the first 6 months of 1988. Exports to Ireland during the period increased by 117 per cent, UK by 28 per cent Portugal by 27 per cent. There was also a marginal increase by 1 per cent in India's exports to the Federal Republic of Germany. However, India's exports recorded negative growth rates in the other countries of the EEC.

The growth rate in the case of saddlery and harness increased by 12 per cent, travel goods and bags by 12 per cent, leather garments by 36 per cent, industrial and fancy gloves by 37 per cent and belts and girdles by 45 per cent. The sole and exception to this was finished footwear, the export of which came down by over 18 per cent.

In the field of joint venture, the EEC occupies an unrivalled position. Out of 926 foreign collaborations approved by the Government of India during 1988, 446 (48 per cent) were the EEC.

(Source: *Hindustan Times*)

LEPROSY

196 districts in the country have a leprosy prevalence rate of over five per

thousand as on end March 1985.

There were 2,58,7564 leprosy patients on record in the Country. Since the inception of the National Leprosy Control Programme in 1955, as many as 43.54 lakh patients have been cured, while 26,794 were medically rehabilitated and 12,983 have been vocationally rehabilitated.

LIFE EXPECTANCY

The life expectancy at birth in India is 58.1 years for males and 59.1 years for females. This was 32.09 years and 31.37 years for male and female respectively in 1941 and 32.45 years and 31.66 for male and female respectively in 1951.

LITERACY

Literacy Rate

The literacy rate in the country has registered an increase by 8.55 per cent during the last decade.

Adult Literacy

Against the target of enrolling 177.3 lakh persons for 1990-91, an enrolment of 118.7 lakh persons have been reported so far. About 2.5 to 3 million volunteers, both student and non-student youth, are imparting literacy to about 25 to 30 million learners through total literacy campaigns which have been launched in 45 districts of the country. The majority of the illiterate population is in the rural areas.

The proposed outlay for adult education and elementary education for 1991-92 is of the order of Rs 178.75 crore and Rs 1036.43 crore respectively, Berwa.

This was against the annual plan allocation of Rs 151.12 crore for adult education and Rs 843.15 crore for elementary education during 1990-91.

Literacy Mission

Rs 96 crore during the 1990 have been provided for the National Literacy Mission. A grant of Rs 2.41 crore has been given to the Bharat Gyan-Vigyan Jatha Samiti for its programme aimed at mass mobilisation for the literacy programme. The Samiti had sought a grant of Rs 25 crore.

AIR Literacy Project

As part of the International Literacy Year celebrations the Department of Education prepared a project entitled "Radio learning package", which aims at imparting literacy through the use of radio in 15 district of Bihar, Madhya Pradesh, Rajasthan and Uttar Pradesh.

The radio learning package would be covered by eight stations of All India Radio involving about 4000 adult education centres for th benefit of about one lakh learners.

Preparatory work has been completed and the proposed project would be ready for broadcast during September 1990, coinciding with the International Literacy Day. The proposed project has been based on the survey of illiterates and training of field functionaries and preparation of software.

The most significant achievement made in India during the International Literacy Year (1990) has been the attainment of the full literacy status by two districts namely Kottayam and Ernakulam in Kerala.

This remarkable achievement in the promotion of adult education and eradication of illiteracy in two districts has encouraged other educationally backward States like Andhra Pradesh, Orissa, Bihar, Uttar Pradesh, Madhya Pradesh and Rajasthan for launching similar literacy drivers.

Goa and the Union Territory of Pondicherry have also taken up a special drive for total eradication of illiteracy to cover about one lakh and five Lakh illiterate people respectively in a period of one year.

(Source: *UNI Release*)

LOANS

As many as 3.03 crore farmers, weavers and artisans have been provided relief of Rs 7,560.88 crore by public sector banks, regional rural banks and cooperative banks by July 1 this year under the Agricultural and Rural Debt Relief Scheme, 1990. Target number of beneficiaries were from Uttar Pradesh (51,63,306) followed by Bihar (37,31,001), Andhra Pradesh (29,51,666) and Maharashtra (28,78,901).

LOTTERIES

The Comptroller and Auditor General of India in the report to the year 1988-89, has drawn attention to the inadequate tax assessment of persons connected with the lottery business, including organising agents and stockists, causing huge loss of revenue to the exchequer.

There were substantial under-charge of taxes of over Rupees six crore, large scale evasion of tax by adoption of questionable modes, and inadequate safeguard against surreptitious sale of winning tickets and dissipation of income, as also commission of centralised assessment of those associated mainly with lottery business for the purpose of coordination.

Absence of one single source of collection and collation of information of lottery assessees so as to widen the tax base, wide fluctuating profits as between assessees and assessment years and absence of evidence of cross-check with tax deduction returns to book liable tax-payers not filing returns have also been mentioned.

There were delays and deficiencies in deduction of tax at source and narrow ambit of compulsory source—deduction of tax to lottery payments and widespread non-compliance of anti-evasion measures as regards compulsory maintenance of accounts, tax audit and payments only through cheques, drafts etc.

There was incorrect exemption in a summary assessment of profits of Rs 60 lakh in the case of a Sports Trust, which had organised two lottery draws without any licence. There was also suppression of sale of tickets by Rs 1.13 cores.

In three other cases though the trusts and become void due to violation of the lottery rules, the income was nevertheless exempted. Further, the sales of tickets were suppressed by Rs 191.51 lakh. Refund of prize money of Rs 25 lakh on unsold tickets was also not accounted for in one case.

On the basis of a post-dated court affidavit the claim of an individual that the prize money of Rs 2.22 crore in a raffle was jointly won with his wife and two sons, was accepted without proper investigation though the affidavit of one of the members had not been accepted.

A registered society, which was permitted to raise funds for repayment of Government loans through lotteries and had earned Rs 405.26 lakh in the business, arranged the conduct of lotteries through another society, agreeing to advance a loan of 30 per cent of the net proceeds of the lottery business for 21 years at a nominal, interest of one per cent. Although the former society had diverted its income for objects other than that of the society, and the service society was also liable to tax proceedings, the department failed to initiate any assessment proceedings on both the societies.

The review of the system by audit disclosed errors and deficiencies in the procedure of collection of tax at source, such as non/short deduction of tax and delays/omissions in remittance of deducted tax involving interest/penalty leviable of Rs 7.48 crore, non-compliance in furnishing of periodical returns by tax payers, coupled with monitoring lapses by the department, with penalty effect of Rs 3.89 crore, non-incomplete maintenance of control registers and non-reconciliation of tax deductions regularly.

No regular reconciliation with bank challans, no visible improvement in administration of 'tax deducted at source' despite creation of exclusive TDS sections, no machinery for effective coordination cross check and verification of inland and foreign deduction certificates slow progress in allotment of Tax Deduction Account Number and absence of internal audit and administrative supervisions were also noticed.

As regards corporation tax, arithmetical mistakes caused due to negligence in the computation of income and tax contributed to under-assessment of Rs 11.21 crore by way of tax in just 679 cases.

M

MADHYA PRADESH EXPORT CORPORATION

The report of the Comptroller and Auditor General of India tabled in the State Assembly cast aspersions on the functioning of the Madhya Pradesh Export Corporation.

Due to the recovery of service charges at lesser rates from the suppliers of soyabean meal extraction on value of exports during 1987-88 to 1989-90 the company suffered a loss of Rs 0.6 crore.

Though the main object of the company is to export goods produced by the medium and small scale industries, its export business of Rs 33.72 crore during the seven years up to 1989-90 constituted only 27.7 per cent of the total turnover of Rs 121.67 crore.

In terms of the agreements entered into with the supply associates for supply of railway sleepers, the company was entitled to service charges at the rate of two per cent. However, the company waived the levy of two per cent service charges amounting to Rs 0.54 crore recoverable from the associates.

MADHYA PRADESH GOVERNMENT

The Comptroller and Auditor General of India has criticised the Madhya Pradesh Government for not having any policy for the procurement, operation and maintenance of State aircraft.

The Government had neither laid down any policy about the type and number of aircraft to be maintained nor had it spelt out the specified consideration for procurement or disposal of aircraft in the relevant proposals.

Non-notification of the change in the type of aircraft offered for sale to all the tenders and negotiations with only one tenderer whose rate was eighth highest, deprived the Government of the benefit of competitive offer.

Besides, the purchaser was provided unauthorised aid of Rs. 3.32 lakh by way of getting one aircraft repaired at Government cost and by handing over additional item (fuel control unit) to him without any consideration.

Disposal of a damaged aeroengine, which could have been repaired, resulted in a loss of Rs 23.40 lakh.

Heavy purchase of spare parts during 1982-88 led to accumulation of huge stocks unlikely to be utilised within next 30-40 years.

MADHYA PRADESH STATE ROADWAYS TRANSPORT CORPORATION

Against the outlay of Rs 82.52 crore to purchase 3,002 new vehicles during the seventh five year Plan and the reduced outlay of Rs 63.30 crore as per annual plans for purchase of 2,052 new vehicles during the same period, the corporation could arrange procurement of 1,648 vehicles only.

MAIN BATTLE TANK (MBT)

MBT project is in an advanced stage of development and a total of Rs 203.71 crore has been spent on it, including the commitments made on March 31, 1991.

So far 12 prototypes of MBT have been fabricated, which are undergoing trials. Order for six pre-production series tanks have also been placed.

MANPOWER

According to a study conducted on behalf of the Department of Science and Technology by the Centre for Research, Planning and Action, there is considerable, present and prospective, imbalance in demand and supply of scientific and technical manpower.

Agricultural scientists, are already in surplus by about 20 per cent. This surplus is expected to rise to about 22 per cent by 1990-91. Thereafter it will have a declining trend and by 2000-01 will decline to 12 per cent.

As against this, shortage of about 1.5 per cent of the demand is observed in case of veterinary scientists and this shortage is expected to increase to nine per cent by 1995-96. And further to 18 per cent by the turn of the century, unless proper corrective steps are taken.

In engineering and technology manpower, a surplus by three per cent is expected by 1990-91 in case of degree holders. This will increase to six per cent by 1990-95 and further to eight per cent by 2000 AD. In case of diploma-holders the surplus will be by nine per cent by 1990-91. Thereafter it will show declining trend and by the turn of the century there will be approximate balance with respect to their demand and supply.

In medical manpower already there was shortage of doctors by about 18 per cent in 1980-81 itself. This shortage is expected to increase to 20 per cent by 1990-91. Thereafter it will short somewhat a declining trend and by 1995-96, the shortage will decline to 17 per cent by 20001 to twelve per cent.

In case of nurses shortage is expected to be substantial. By 1990-91 it will be by 39 per cent of the demand and this shortage will further rise to 50 per cent by 2000 AD.

Substantial shortage is also anticipated in case of health educators, lab-technicians, nurses-midwives, health assistant (male) and health workers (male).

Gaps in supply in relation to demand are also expected in some emerging areas. A large number of new installations that are being established in the area of electronics and telecommunications will be in need for properly trained manpower to maintain and keep the industry productive.

In the energy sector, for new technologies which have been developed to meet the need of the rural sector sufficient manpower or good talent is needed to meet the requirements. Also in food processing, textiles designing, printing and computer science additional technical manpower of adequate capabilities are required.

By the year 2001 there will be a shortage of 385,000 in medical and nursing group, about 52,000 in engineering and technology group and a surplus of 13,700 in agriculture and veterinary group and 303,000 in general science group.

For 1990-95 a surplus of 342,000 in general science group 28,000 in agriculture and veterinary group and a shortage of about 3,000 in engineering and technology group and 255,000 in medical and nursing group is estimated.

(Source: *UNI Release*)

MASSACRES VICTIMS

See also Punjab Killings; ULFA Killings

Massacres Victims

Year	Place	Number of persons killed
1977	Belchi	14
1980	Pipra	14
1980	Paras bigha	11
1984	Gagan bigha	05
1986	Gaini	12
1986	Kansara	11
1988	Nonhi-Nagwan	18
1988	Koiria-Chatar	07
1989	Munger	07
1989	Danwar Bihta (Bhoipur District, Bihar)	23
1990	Dariapur	04
1991	Paharpur	07
1991	Harlakhi	06
1991	Mabaria	13
1991	Deo and Sahiara (Bhoipur district, Bihar)	14

Source: *India today,* July 31, 1991.

MARUTI

The Maruti Udyog earned Rs 72.91 crore in foreign exchange during 1990-91 from the sale of vehicles, but it imported components worth Rs 171.14 crore in foreign exchange.

The company exported 4,908 vehicles, including 3,187 Maruti-800s and 1,502 Gypsies. It produced 79,226 cars during the year.

MEHAM PROBE

An expenditure of Rs 6,72,871 was incurred on the Meham Inquiry Commission During 142 days last year on facilities like office accommodation, staff funds, bungalow, transport and other arrangements necessary to run an office which were made available to the Commission.

MERCHANT SHIPS

In 1989 orders for construction of seven merchant ships had been placed with foreign shipyards. No orders were placed for construction of merchant ships during 1990-91.

MIGRANTS AND REFUGEES

Rs 5 crore has so far been released to the Government of Jammu and Kashmir towards expenditure on relief assistance to migrant families.

See also Bangla Desh Refugees; Kashmir Situation; Sri Lankan Refugees.

MILITANTS

See Kashmir Situation; Punjab Killings; Sri Lankan Refugees; United Liberation Front of Assam.

MILITARY AIRCRAFT

The Comptroller and Auditor General of India has blamed the Ministry of Defence for inducting a military aircraft into the Indian Air Force without deciding what kind of weapons systems were required for it.

There was a mismatch between the induction of the aircraft into squadron service and that of the weapons required for it.

While the contract for the aircraft was concluded in October 1982, the contracts for weapons were finalised after two to three and half years resulting in inordinate delay in the availability of the weapons.

Owing to the delay in the availability of the weapons required for training and instructional purposes, the Indian Air Force (IAF) depended largely upon simulation methods and dummy modes. A single supplier situation prevailed throughout in respect of all weapons.

Price escalation amounting to Rs 57.39 lakh in foreign currency had to be paid due to delay in finalisation of contracts.

An incorrect assessment of the quantitative requirement of a weapon initially and the procurement of an additional quantity subsequently at a higher rate led to an avoidable extra expenditure of Rs 1.12 crore.

Owing to the absence of a specific clause in the agreement regarding the period for which the aircraft will be kept with the manufacturer for weapon integration and clearance, four aircraft remained with the manufacturers for long periods leading to deficiency of air craft in the squadrons.

In 1983 the Government initially approved procurement of various types of weapons systems at an approximate cost of Rs 261 crore with an option to order the balance of weapons worth Rs 28 crore in due course. These weapons were ordered when they were still under development with the foreign firm. The squadrons were raised in June 1983 and January 1986 without weapons for instructional, purposes.

MINERALS AND METAL TRADING CORPORATIONS (MMTC)

MMTC has 1989-90 a record turnover of Rs 4,961 crore in 1989-90. While much of this increase is due to higher imports of fertilisers, for the first time, its exports have crossed the Rs 1,000 mark.

As against Rs 873 crore the previous year, the MMTC exports in 1989-90 are estimated at Rs 1,801 crore. The rise has been more pronounced in the case of canalised items. Iron ore exports, for instance, crossed Rs 500 crore, while manganese ore exports at Rs 36 crore is also an all-time high.

In the non-canalised category, a significant jump has been noticed in the export of gold jewellery, diamonds and gem stones — from Rs 60 crore the previous year to Rs 127 crore. Export of Tata buses to Togo and Senegal, earth moving equipment to Jordan and railway equipment to Iraq have also contributed to the better performance.

Counter-trade, has been adopted as one of the main thrusts for higher export turnover.

The MMTC has been nominated to organise counter-trade exports for a total value of US $ 200 million against the hydro projects of the National Hydro Power Corporation (NHPC). In the purchase of DAP fertiliser, also, producers and suppliers have been persuaded to accept counter-trade from India.

MMTC had a plan to set up three joint project — two in Jordan and one in the US — for manufacturing DAP and phosphoric acid

Rock phosphate, was available in plenty in Jordan. The projects to be set up there plan to exploit this potential. The US projects aims at purchasing rock phosphate and converting it into phosphoric acid, which could then be imported to India on better terms for manufacturing phosphatic fertilises.

The export target for 1990-91 had been fixed at Rs 1,221 crore (Rs 671 core for canalised items and Rs 550 crore for non-canalised items). The main emphasis now had to be on export of non-canalised items and counter-trade exports.

If the canalised exports are unlikely to grow this year, it is mainly because of the reduced export of iron ore to Japan since ore availability from Bailadila will be lower. With the Vizag steel plant commencing production, the Bailadila ore will be required for its use. A number of pig-iron and sponge iron plants, now being set up, are also going to require iron ore, leaving relatively smaller quantities of exports.

In respect of manganese and chrome ores, too, the quantitative ceiling on their exports, imposed from the conservation point of view, will be impediment in steeping up their exports. The demand for chrome ore, is declining at the moment in the world market.

On the imports side, fertilisers have accounted for the steep rise in the MMTC imports during 1989-90. Fertiliser imports during the year and shot up to Rs 1,975 crore from Rs 1,072 crore the previous year. Rock phosphate imports were around Rs 30 crore.

In order to minimise the outgo of foreign exchange, the MMTC stepped up its imports from rupee payment areas during the year. These imports were of the order of Rs 543 crore as against Rs 336 crore the previous year.

Jordan has become a regular market for export of earth moving equipment, manufactured by the Bharat Earth Movers Limited (BEML). Last year, repeat orders worth Rs 7 crore were obtained for these equipment.

MMTC had also achieved a breakthrough in exporting frozen meat to Jordan. Such export from India to Jordan was banned earlier. The MMTC took the initiative to get the meat tested by the Jordanian authorities. Therefore, Jordan lifted the ban. The export of this item in the current fiscal year could go up to Rs 10 crore.

(Source: *Hindustan Times*)

MINES AND MINING

All the Public Sector Undertakings under the Department of Mines with the sole exception of Bharat Gold Mines Limited earned profits in 1989-90. HZL and National Aluminium Company Limited (NALCO) earned their highest ever profits. The comparative figures of turnover and profits in 1989-90 and 1988-89 are given in Table.

Turnover and Profits of PSE

(Rs in crore)

	1988-89		1989-90	
	Turnover	Profit	Turnover	Profit
NALCO	438.75	12.96	859.91	120.95
BALCO	356.22	12.65	418.61	3.00
HCL	430.65	46.64	460.90	45.15
HZL	284.30	35.00	415.00	85.19
MECL	53.33	1.74	52.15	0.15

Bharat Gold Mines Limited continued to incur losses due to the uneconomical mining of gold at depths of more than 300 metres. The Company is now diversifying into undertaking projects and contracts for other companies and manufacturing mining equipment. It increased its turnover in these areas in 1989-90.

Wage Agreements

The wage agreements of three companies — Mineral Exploration Corporation Limited (MECL), Bharat Gold Mines Limited (BGML), and National Aluminium Company Limited (NALCO) were finalised after approval by the Government. It expected that a period of industrial amity and harmony will prevail resulting in greater productivity.

Continuous Cast Copper Wire Roads Plant

The Continuous Cast Copper Wire Plant at Taloja, Maharashtra was dedicated to the nation in March 1990. The plant is presently producing 3000 tonnes per month resulting in a foreign exchange saving of approximately one crore Rupees per month.

The plant has an installed capacity of 60,000 tonnes per annum. It will substantially meet the country's demand for Continuous Cast Copper Wire Rods which are being imported at present. At full capacity utilisation, it will save approximately Rs 20 crores per annum in foreign exchange.

Implementation of Integrated Lead-zinc Project of HZL

The Rs 617 crores Rampura-Agucha-Chanderiya Integrated Lead-Zinc Project was sanctioned in November, 1988. The project, when fully commissioned, is expected to produce 70,000 tones of Zinc and 35,000 tonnes of Lead, thereby reducing dependence on the imports of these metals substantially. The rigid schedule prescribed for the timely completion of the project is being maintained. The project is expected to be completed by March, 1991.

Production and Exports of Aluminium and Alumina

The highest ever production of 4.27 lakh for Aluminium was achieved in 1989-90 as compared to 3.57 lakhs tonnes in 1988-89. NALCO produced 1.35 lakh tonnes in 1989-90 which was almost double as compared to 0.78 lakh tonnes in 1988-89.

A major thrust has been given to exports. During 1989-90 Rs 491 crores of Alumina and Aluminium were exported as compared to Rs 27 crores in 1988-89.

Import Policy Changes

In view of the easy availability of aluminium, the imports of aluminium including aluminium alloys an aluminium scrap/dross were removed from the OGL list and brought on the list of restricted imports. This will reduce unnecessary impoorts of Aluminium, thereby conserving foreign exchange.

The import of copper cathodes will no longer be canalised. The import of tin has also been decanalised. It is expected that these measures wire result in more cost effective import by the consumers themselves

Geological Mapping

A total of 2.75 million sq.km (about 17%) of the approachable area of the country has been covered by Geological Survey of India (GSI) by systematic geological mapping. This compares with even advanced countries. The GSI has also covered 37% of the targeted areas of the country by National Aeromagnetic Surveys, and 56% of the Exclusive Economic Zone (EEZ) through marine surveys.

Mine Workers' Allowance

The Government has revised the special allowance for mine workers from April 1, 1991. The minimum wage for unskilled workers will increase from Rs. 17.88 to Rs. 19.34 per day for workers engaged above ground and for those working under ground will be Rs. 23.52 as against Rs. 21.69.

The revised wages for semi-skilled and unskilled supervisory staff will be Rs. 24.12 as against Rs. 22.29 at present for above ground work. The wages for skilled and clerical workers will be Rs. 29.89 against Rs. 27.62 at present.

For the workers engaged under ground, the rate of wages per day for semi-skilled and supervisory staff will be Rs. 28.91 as against Rs. 26.72 per day and for those in skilled category the wages per day will be Rs. 35.72 as against Rs. 32.92.

The mines covered are : red oxide, quartz quartzite, silica, magnesite, granite, laterite, stone, kyanite, steatite (including mines producing soap stone and talc) orchre, asbestos and fire clay, copper, uranium, iron ore, hematite, rock phosphate, chromite, magnetite, gypsum and barytes, graphite, dolomite, china clay, white clay, worfram, manganese felspar bauxite, marble and calcite and mica mines.

(Source: *Yojana*)

MINORITIES

The percentage of Hindus under the last Census in the country was 82.34, Muslims about 11.35, Christians just over 2, Sikhs just over 1, Buddhists 0.71 and Parsis 0.01.

Hindus were in minority in Kashmir

(34 per cent), in Meghalaya (14 per cent) and in Punjab (36 per cent), but the benefits under the 15-point programme were not available to them.

Although the Constitution did not identify the minority communities, the five minorities which were identified at the national level were Muslims, Christians, Sikhs, Buddhists and Zorastrians. The Government had not accepted the concept of the minorities at the State level.

MISSILE

India's most modern medium-range surface-to-air missile, *Akash*, with multi-target handling capacity was successfully test-fired. *Akash* is the fifth Indian-made missile.

The only comparable system to *Akash*, which has a 25-km range, is the US Patriot.

Before test-firing of *Akash*, the phased array radar, which has 2,000 transmitting elements, had been tested. The full system would have 4,000 elements. The technology for this radar would have applications in the Indian airborne surveillance and warning control system (AWACS).

Akash belongs to a family of five missiles developed by the Defence Research and Development Organisation (DRDO). The other four — *Agni, Prithvi, Trishul and Nag* — have already been tested successfully. *Prithvi*, the surface-to-surface, and *Trishul*, the multi-role quick reaction short range surface-to-air missile, are already in production and will be inducted into service by 1992. *Trishul* has also undergone successful trials as anti-skimming role for the navy.

(Source: *Hindustan Times*)

MISSILE BOAT

The first missile boat named 'Vibhuti' to be built in India was launched at the public sector Mazgaon Dock (MDL).

The boat is being built for the Indian Navy, in collaboration with the Soviet Union. She is slated to joint the naval fleet by middle of the next year, after testing, tuning And trials.

Government had approved acquisition of 20 missile boats by outright purchase from the USSR and indigenously construct another nine missile boats at Mazgaon Ddocks and six at Goa Shipyard in coming years. 70 per cent of the equipment have been already installed inside the 'Vibhuti'.

MISSIONARIES

A total of 1,897 foreign missionaries were in India as on Jan 1, 1989, with the highest number of 278 from the USA.

The number of missionaries from Britain was 182 followed by Italy 177, Germany 144, Spain 124, France 122 and Ireland 111. One missionary each was from Guinea, Hungary, Zaire, Jordan, Romania and South Africa.

MONEY SUPPLY

In an occasional paper on Trends in Prices, brought out by Federation of Indian Chambers of Commerce and Industry (FICCI), it is maintained that a major factor causing pressure on prices is increase n money supply over and above the increase in production. For instance, in the past 10 years, the compound rate of growth in respect of money supply was 14.4 per cent per year. The compound rate of growth of production was only 4.9 per cent per year, and the resultant rate of inflation was 8.9 per cent per year. Price rise upto a certain level, say five per cent, is manageable in a developing country like India. But rate of increase of prices over this limit is a matter of concern.

Stability in the supply side is a precondition for maintaining a stable price in the long run. However, nearly one-third of GNP comes from agriculture which is susceptible to wide fluctuations. Besides, agricultural products influence cost of living and consequently wage and cost of industrial production and, therefore, prices.

A one per cent increase in the prices should raise industrial cost by 0.25 per cent. Also, it would result in transfer of income from urban to rural areas. Taking into account the marketable surplus of foodgrains and the amount of cash crops directly sold to industry, the rise in prices, in the first quarter of 1990-91 should transfer Rs 18,000 crores to the rural sector, enhancing their purchasing power.

Therefore, the safe way to maintain the price level is to increase the production of agriculture, through introduction of biotechnology and genetic engineering. Particular attention shod be given to production of pulses and oilseeds which have always fallen short of demand and consequently become victims of inflation and imposed burden on balance of payments.

Equally important would be improving the productivity of industry sector. All countries which sought to linkwage to cost of living were unable to control prices and therefore wages should be linked to productivity. Increase in the prices of administered commodities like coal, steel, etc., also had a tremendous spread effect. In the last two years, administered prices of steel, aluminium and coal prices raised quite periodically, contributing to higher inflation.

A major factor which has pushed up prices of items like wheat, barley, etc., was the enhancement of procurement. minimum support prices. Procurement prices of wheat to be marketed in 1990-91 has been raised to Rs 215 per quintal from Rs 183 per quintal. Similarly, minimum support prices of barley of fair and average quality has been increased to Rs 180 per quintal from Rs 145 per quintal.

Prices of pulses were under pressure mainly because of imbalance between demand and supply. The increase in production has not kept pace with the rise in population. Whereas population has been rising at over two per cent per year, production of pulses went up only marginally from 13.36 million tones in 1985-86 to 13.7 million tonnes in 1988-89. To increase the availability of pulses, imports have gradually gone up and touched 8.26 lakh tonnes in 1988-89. Despite domestic production being supplemented by imports, the total availability of pulses fell short of requirements. The per capita availability of pulses has declined to 40.4 grams in 1989 from 41.9 grams in 1986.

Prices of edible oils rose in the current financial year partly because of decline in their production to 46 lakh tonnes in 1989-90 from 50.5 lakh tonnes in 1989-90. The other factor responsible for rise in price of edible oils is the vast difference between the price at which edible oils are imported and the price at which these are supplied to vanaspati manufacturing units.

The cause for rise in price of cement was largely due to substantial increase in demand on the one hand and shortage of wagons to move cement from factory to market place on the other. Cement price has also been affected by increase in input cost.

(Source: *Hindustan Times*)

Money Supply with Public

(Rs. in crores)

Year	Money Supply
1984	33398
1985	39915
1980	44095
1987	51516
1988	58555
1989	71099
1990	81608
1990 (June)	86273

MONOPOLIES AND RESTRICTIVE TRADE PRACTICES (MRTP) ACT

The number of companies under the purview of the Monopolies and Restrictive Trade Practices (MRTP) Act has more than doubled during the last 17 years of the operation of the Act.

The number of undertakings has gone up from 825 on January 2, 9172 to 1, 816 on December 31, 1989.

Substantive increase has taken place during 1975 and 1984 when definition of "interconnection/same management" was changed. Subsequently, the number of registered undertakings came down to 1,652 by December 31, 1986 as a result of an amendment when the asset limit was raised from Rs 20 crore to Rs 100 crore. Recently against there was an increase in the number of registrations.

Between May 29, 1985 when the asset limit was raised and till December 31, 1989, the Department of Company Affairs has received 848 applications for registration. Of these, 472 had been approved and 365 cases had been rejected as the undertakings were found either interconnected with registered undertakings or because they were dominant. The remaining 11 applications are pending, according to official sources.

Assets of the large houses have also increased each year and were of the order of Rs 44,560 crore in 1987-88. The corporate sector has shown a remarkable growth since coming into force of the Companies Act, 1956. The number of companies at work increased by about seven-fold from just 29,357 as on March 31, 1957 to 1,92,999 as on December 21, 1989. The paid-up capital also increased by as much as 50.6 times from Rs 1,078 crore to Rs 54,500 crore during the same period.

The growth during the last five years ending December 31, 1989 was of the order of 88 per cent (17.6 per cent annual) in terms of number of companies at work and 97 per cent (19.4 per cent annual) in terms of their paid up capital.

Government companies had also recorded a faster growth rate than the non-Government companies. The number of Government Companies increased from 74 as on March 31, 1957 to 1,155 as on December 31, 1989 or by more than 15.6 times as against 6.6 times increase in the number of non-Government companies.

The paid-up capital of Government companies increased phenomenally by 544 times from only Rs 73 crore as on March 31, 1967 to Rs 42,443 crore as on December 31, 1989.

The paid-up capital of non-Government companies increased only by 12 times during the period. As a result, the shares of Government companies in the paid-up capital of corporate sector increased from only 7 per cent in 1957 to 77.9 per cent as on December 31, 1989.

(Source: *Hindustan Times*)

MORTALITY RATE

Nearly 11 people die against every 1,000 births in India. The higherst mortality rate in India is in Arunachal Pradesh where it is 17.0 against the national average of 10.9 per thousand. The Union Territory of Chandigarh has the lowest mortality rate of 4.5 per thousand.

The estimated death rate for the decade 1941-51 was 27.4 per thousand. Neonatal mortality rate for the country is 57.5 per thousand live births.

MULTISTAGE FLASH (MSF)

India has commissioned its first multistage flash (MSF) desalination plant at Trombay. Designed by engineers at the Bhabha Atomic Research Centre (BARC), the plant has a capacity to desalinate up to 425 cubic metres of sea water per day. It represents the prototype of larger commercial size desalination facilities.

The MSF process accounts for about 70% of the world's desalination capacity, most of it located in the Middle East region. The potential of India to compete in international markets for MSF plants is considered good, since the manufacturing technology is well established in the country. All the equipment and most materials for the Trombay desalination plant were domestically produced and supplied.

(Source: *Nuclear India*)

MUNICIPAL CORPORATION OF DELHI

The Comptroller and Auditor-General of India (CAG) has roundly criticised the Municipal Corporation of Delhi (MCD) for excessive spending over income in its general wing, and other lapses in the corporation's water supply and sewage disposal undertaking and electric supply undertaking.

The provisional figure for 1988-89 showed that the general wing of the corporation had spent Rs 340.32 crore against an income of Rs 330.46 crore. The revenue expenditure of Delhi Electric Supply Undertaking (DESU) was Rs 572.30 crore against receipts of rs 419.63 crore, leaving a revenue deficit of Rs 152.67 crore. The capital expenditure was Rs 190.09 crore against receipts of Rs 157.58 crore.

The expenditure of Delhi Water Supply and Sewage Disposal Undertaking (DWSSDU) was Rs 162.56 crore, against receipt of Rs 99.94 crore, leaving a deficit of Rs 62.62 crore.

The general wing of the corporation did not submitted the annual appropriation accounts from 1985-86 and the annual accounts for 1988-89 to the chief auditor. The DESU had not submitted the accounts from 1985-86 and annual appropriation accounts from 1981-82.

Temporary advances to the extent of Rs 540.07 lakh paid to the supplies during the 1954-55 to 1988-89 had not been adjusted till end-March 1989. Vouchers for Rs 1,229.49 lakh and payees stamped receipts aggregating Rs 52.01 lakh were outstanding in accounts up to 1986-87, which included large sums drawn more than a decade earlier.

No internal audit manual had been compiled so far. Internal audit of the general wing had neither conducted the internal audit of all the units nor had it taken proper follow-up action for the settlement of outstanding objections.

The Corporation incurred expenditure for construction of multi-tiered stages, providing stage lighting, sound arrangement for the culture festival held in Delhi from November 8, 1986 at 27 venues costing Rs 131.36 lakh, against which Rs 75 lakh was received from the national cultural festival so far. The NCF not accepted the liability for the excess amount incurred by the corporation.

An expenditure of Rs 49.79 lakh was incurred on hiring of electrical equipment including lighting and the public address system, against the original estimate of Rs 8 lakh.

Medicines valuing Rs 9.70 lakh, had been accepted in 1987-89 in contravention of the conditions of supply as more than one-sixth of life of the medicines had expired.

Arrears of property tax were Rs 3,187 lakh at the end of 1984-85 and went to Rs 3,673 lakh at the end of 1987-88.

No manual had been compiled laying down the procedure to be followed and records to be maintained in respect of assessment, raising of demand, recovery and accountal of property tax. There was also no coordination with the engineering department of the corporation which approves the building plans and the Delhi Development Authority which is a major agency for building and selling properties in Delhi.

Pass-books meant for issue to assessees of property tax printed in July 1987 at a cost of Rs 1.12 lakh had not been issued. New Delhi Municipal committee had not paid its share of fire tax since 1958 and this had piled up to a colossal Rs 650 lakh by 1987-88.

MUSEUMS

There are about 400 museums in the country and grant in-aid are given by the Central and State Governments from time to time.

Funds by the Union Department of Culture are given to following five museums: National Museum (Delhi) Indian Museum and Victoria Memorial Museum (both Calcutta), Salar Jung Museum (Hyderabad) and Allahabad Museum.

N

NARMADA ACTION PLAN

Action plans are being implemented for cleaning the Ganga, Narmada and Kshipra rivers. While the Ganga Action Plan is being implemented with full grant from the Centre, the Narmada Action Plan is being implemented by the Government of Madhya Pradesh and this included the cleaning of Kshipra. Of the 52 schemes sanctioned for Narmada and 11 for the Kshipra, works had been completed on 29 and 3 schemes respectively.

See also Ganga Action Plan.

NATIONAL CAPITAL REGION PLAN

The National Capital Region Planning Board has envisaged an investment of Rs 2,900 crore during the Eighth Plan period (1990-95). In addition, Rs 828 crore has been proposed for the five counter-magnet areas outside the NCR.

The allocation for the Plan's implementation during 1990-91 has been a meagre Rs 10 crore.

The NCR Planning Board had sought Rs 800 crore for the NCR projects during the Seventh Plan. Against that, the Government released only Rs 60 crore for the entire five-year period.

A "Regional Plan—2001" was finalised for the NCR and made "effective" from January 1989.

Central allocation for the first year of the Eighth Plan has been just Rs 10 crore. Although the Board is a statutory body which came into being in March, 1985, the recommendations of the Board have been merely advisory as far as their financial implications are concerned.

The investment plan for "Regional Plan—2001" envisages an investment of Rs 1750 crore in the Central sector and Rs 1150 crore in the State sector. The counter-magnets have been "provided" a sum of Rs 828 crore of which Rs 728 crore are to be in the Central sector.

Unless Delhi's population growth is checked, it is expected to be 132 lakh in another 10 years. The Regional Plan envisages integrated development of self-contained priority towns and the Delhi Metropolitan Area (DMA) towns and counter-magnets outside the region as interceptors of potential migrants to Delhi.

The DMA towns which are proposed to hold a population of 37 lakh are Ghaziabad, Loni, NOIDA, Faridabad, Gurgaon, Bahadurgarh and Kundli towns. The priority towns are Meerut, Hapur, Bullandshahr Khurja, Panipat, Rohtak, Palwal, Alwar and Rewari-Dharuhera-Bhiwadi complex.

The suggested outlay for the Central sector includes Rs 219 crore for national highways, Rs 45 crore for expressways, Rs 414 crore for railways, Rs 375 crore for telecommunications and Rs 697 crore for power development.

The States sector outlay includes Rs 460 crore for land acquisition and development for residential use , Rs 187 crore for land acquisition and development of economic activities including industrial development, wholesale trade, commercial complexes office-accommodation.

(Source: *Hindustan Times*)

NATIONAL COUNCIL ON SCIENCE AND TECHNOLOGY

The Union Cabinet gave approval on June 7, 1990 for the establishment of the Council to be headed by the Prime Minister. The 18 members include the Deputy Prime Minister, Finance Minister, Deputy Chairman of the Planning Commission, Minister of Human Resource Development, Minister of State in charge of Science and Technology, Defence and Education.

The Council is expected to provide policy and directions for development and applications of science and technology in the country.

NATIONAL FERTILIZERS LIMITED (NFL)

NFL, the largest producer of

Nitrogenous fertilizers in the Public Sector, has exceeded the production target for the year 1990-91, achieving an overall average capacity utilisation of 98.2 per cent. The plant at Vijaipur achieved an overall capacity utilisation of 117 per cent during the year breaking all previous records. The Company also achieved the highest sales during the year 1990-91, exceeding the Rs. 1000 crore mark. This represents a 30 per cent growth over the previous year. This has resulted in substantial reduction in the inventory which was being carried by the Company in the last 2-3 years due to the glut prevailing in the industry.

NATIONAL HYDRO-ELECTRIC POWER CORPORATION (NHPC)

The public sector NHPC took up three projects in Jammu & Kashmir, namely Uri (480 MW), Dul Hasti (390 MW) and Salal Stage II (345 MW) and one each in Uttar Pradesh, Himachal Pradesh Chamera Stage 1 (540 MW) H.P. Tanakpur (120 MW) U.P. and Sikkim, Rangit (60 MW). It has, in addition, embarked on construction of electric towers and laying of transmission systems. During 1990-91, the operating unit of NHPC generated 3617 million units against the target of 3198 m.u., marking a rise of 113 per cent. Capacity utilisation during the period at 112 per cent was higher by 6 per cent compared to the previous year.

The transmission from Chukha hydro-electric power projects in Bhutan at 1374 million unit was also high compared to the earlier year. The Corporation made a profit of Rs 52.71 crore, which was marginally higher than in 1989-90. With its proven experience in operation of transmission system, the Corporation has bagged a prestigious project — $ 485 million World Bank—aided Northern Regional Transmission System. In the field of tower erection, NHPC has exceeded its target set for 1990-91. As against 507, towers erected during the period were 622.

Among the trend setting activities, the Corporation has, for the first time in the country, started tunnel boring work with machine having a diameter of 8.3 metres for diverting Chenab river water at Dul Hasti project. It has also undertaken a work which involves laying of special coating on underwater parts of the installation at Baira Siul (H.P.) plant to withstand heavy silt erosion. The Corporation has bagged the Indira Gandhi Memorial Award for excellence in public sector.

NATIONAL MINERAL DEVELOPMENT CORPORATION (NMDC)

The Comptroller and Auditor General of India has flayed for colossal cost escalation in its Bailadila from Ore Project and poor project implementation with resultant deficiencies in several stages.

Though the company was registered in 1958, it has not formulated its long-term and micro objectives as yet. Bailadila Iron Project-5, scheduled to be completed in January 1974 was commissioned in January 1979, incurred an expenditure of Rs 67.49 crore, against the original estimate of Rs 36.53 crore. The project never attained the designed capacity since inception. Bailadila Iron Ore Project-14 was provided with two lines of crusher. The detailed project report of Bailadila Iron Ore Project 11-C, accordingly, contained the provision of transfer the spare line of crusher from Bailadila 14. Contrary to this a new line of crusher was assembled by utilising spars from Bailadila-14 by additional procurement worth Rs 389.51 lakh which was charged to Bailadila-14.

Donimalai Iron Ore Project, originally estimated to cost Rs 19.46 crore in 1971, was commissioned in 1977 at a cost of Rs 41.18 crore.

This Donimalai project was never operated at its rated capacity due to absence of marketing tie-up for sale and the matching infrastructure facilities. As a result, the huge investment deployed in the project could not be economically made use of.

The percentage of utilisation of equipment was less than the norms fixed by Bureau of Industrial Costs and Prices for assessing the standard cost price at Bailadila Iron Ore Project-14, Bailadila Iron Ore Project-5 as well as Donimalai Iron Ore Project.

The actual production of fines ranged between 3.58 to 15.81 lakh tonnes per annum, whereas the sales ranged between 1.41-11.42 lakh tonnes during 1978-79 to 1987-88.

The quantity of fines not sold due to inadequate markets was dumped into a valley. The quantity so dumped up to March 31, 1988 was 11.19 lakh tones. the production of lump ore was only about 2 million tonnes per annum, against the target of 4 million tonnes at Bailadila iron ore project.

The corporation look up Diamond Mining Project which included Ramkheria and Majhgawan mines. The company closed the mine in June 1980 and the loss suffered during 1973-74 to 1980-81 amounted to Rs 158.87 lakh.

Though an expansion scheme taken up at Majhgawan mines with a view to increasing the income by Rs 14.80 lakh per annum was completed in November 1982 at a cost of Rs 55 lakh, the diamond mining project has been incurring a loss of more than Rs one crore every year even after completion of expansion.

A scheme to develop adjoining deposit 11-C at an estimated cost of Rs 9.90 crore, however, cost Rs 30.89 crore up to March 1988, in addition to Rs 4.04 crore being spent on installation of a second crusher line. At the end of March 1988, the company showed a loss of Rs 1,527.85 lakh, after wiping off the cumulative profit of Rs 255.73 lakh at the beginning of the year.

NATIONAL SECURITY ACT (NSA)

Between December 1, 1989 and end June 1990 year, as many as 427 persons were detained under NSA with the largest number in Maharashtra at 155, followed by Madhya Pradesh at 99.

NATIONAL TEXTILES CORPORATION (NTC)

NTC Mills incurred a net loss amounting to Rs 206.66 crore in 1989-90

This was against a loss figure of Rs 311.66 crore in 1988-89. This represented a reduction in loss of about 34 per cent.

NATURAL CALAMITIES

At least 2,209 people lost their lives and 312 lakh people were affected in natural calamities in 1990-91. Over 18.7 lakh houses were damaged in calamities and 26.9 lakh hectares of crops are damaged.

NAVAL SHIP

Indian Naval Ship (INS) *Kuthar* the second missile corvette of the *Khukri* class became a man-of-war hoisting the Indian naval engine and raising a "dressing line" of signal flags on board. INS *Kuthar* in a 1300-tonne displacement 90 meter long ship armed with four surface-to-surface missiles, one rapid firing medium range twin-barrel AK-176 gun for both surface and air targets, two AK-630 guns for close range targets and also surface to air missiles.

NAVAL STRENGTH

Comparative Naval Strenghts

	Major ships: coastline (Per 100,000 km coast)	Major ships: Exclusive Economic Zone (EEZ) (Per 100,000 km EEZ)
China	1.0	1.27
India	**0.4**	**0.164**
Japan	0.4	0.168
Pakistan	2.0	0.72

(Source: *India Today* July 31, 1991)

NAXALITES

Andhra Pradesh and Bihar have witnessed a spurt of Naxalite activities in recent years. While the extremists be longing to the People's War Group have been active in Andhra Pradesh and bordering districts of Maharashtra, Madhya Pradesh and Orissa, in Bihar activities of the Naxalite groups are concentrated mainly in Palamu, Gaya, Jehanabad, Patna, Nalanda and Hazaribagh district, he said.

A total of 303 incidents of violence involving Left wing extremists were reported between December 1989 and March 1990 from the four most affected States of Andhra Pradesh, Bihar, Maharashtra and Madhya Pradesh in which 96 people were killed.

NEWSPRINT

The requirement of newsprint in the country was being met, both by indigenous production and imports.

Production of newspring by all the newsprint mills in 1989-90 was 2.64 lakh tonnes, against 2.76 lakh tonnes in 1988-89.

The indigenous production of newsprint is not adequate to fully meet the requirements at present.

Even though sufficient capacity has been sanctioned by way of industrial licences and letters of intent in addition to the existing installed capacity of 3.13 lakh tonnes, the progress of its implementation in the private sector had not been satisfactory, mainly on account of high capital cost and low profits.

91152 tonnes of newsprint worth Rs 10044.48 lakh was imported during 1989-90 from Rupee Payment Area (RPA) countries.

During 1987-88 it was 76881 tonnes worth Rs 7265.85 lakh while in 1988-89 it was 96095 tonnes worth Rs 6915.37 lakh.

The increase in prices of imported newsprint ranged from 22 to 25 per cent.

The total demand of newsprint during 1991-92 is estimated at 5.85 lakh tonnes. The indigenous production of newsprint during the year is estimated at 2.95 lakh tonnes. the shortfall is to be met by imports, subject to availability of foreign exchange.

NEYVELI LIGNITE CORPORATION

The mines, power stations and fertilizer plants of Neyveli Lignite Corporation have been achieving more than the production targets during the last three years. Production in mines in 1989-90 was about 107 per cent of the target. In respect of power generation and fertilizers, the output was about 107 per cent and 111 per cent . NLC has proved that given the required inputs, support and encouragement, public sector unit can perform better than private sector enterprises.

(Source: *Yojana*, December 1-15, 1990)

NON-ALIGNED NATIONS

NAM Centre

Thirty-two countries have so far joined the Centre for Science and Technology of the non-aligned and other developing countries set up at New Delhi about a year and half ago.

NON-PLAN EXPENDITURE

An analysis of the study on Budget highlights 1979-80 to 1990-91 conducted by Confederation of Engineering Industry (CEI) reveals massive increase in the Non-Plan expenditure over the 10-year period.

Receipts of the Government has increased substantially over the years from Rs 15,410 crore in 1979-80 to Rs 75,946 crore in 1989-90, recording an increase of 17.3 per cent compound per annum. Revenue receipts aggregated Rs 51,341 crore representing 67.6 per cent while capital receipts aggregated Rs 24.605 crore accounting for 32.4 per cent. Tax revenue accounted for 79 per cent while non-tax revenue shared 21.0 per cent in 1989-90 as against 85.6 per cent and 14.4 per cent respectively in 1979-80. In 1990-91 the share of tax revenue is projected to increase to 83.5 per cent and that of non-tax revenue 16.5 per cent.

Growth of tax revenue has surpassed the growth of GNP over the period under review. Consequently, the share of tax revenue in GNP has increased from 10.7 per cent in 1980-81 to 12.8 per cent in 1988-90. Statistics show while GNP recorded in increase of 13.9 per cent compound per annum, gross tax revenue shot up by 16.5 per cent compound per annum during the same period.

There is declining share of direct tax in the total tax revenue. Indirect tax, which in a way affects every strata of society, forms the bulk of the tax revenue accounting for 80.1 per cent in 1989-90 as against 75.9 per cent in 1979-80. The share of direct tax has correspondingly decreased from 24.1 per cent in 1979-80 to 19.9 per cent in 1989-90.

An analysis of the composition of tax revenue suggests corporate tax and income tax from the bulk and the two together account for 96.6 per cent of the direct tax revenue. Corporate tax accounts for 46.8 per cent in 1989-90 and is expected to increase to 50.8 per cent in 1990-91. The share of income tax is 49.2 per cent in 1989-90 and is estimated to decline to 45.3 per cent in 1990-91.

Among the components of indirect tax, excise duty formed the bulk accounting for 54 per cent and is expected to account for 52.6 per cent in 1990-91. However, as compared to 1980-81 the share has dropped from 64.2 per cent. Customs duty on the other hand has increased its share to 43.7 per cent in 1989-90 and is projected to improve further to 44.9 per cent in 1990-91 as against 33.6 per cent in 1980-81.

The non-plan expenditure increased by 18.8 per cent compound per annum during the 10-year period thus recording in increase on its share in total expenditure from 59.6 per cent in 1979-80 to 67.5 per cent in 1989-90. The share of non-Plan expenditure is expected to increase further to 68 per cent in 1990-91.

(Source: *Hindustan Times*)

NON-PROLIFERATION TREATY (NPT)

India did not sign NTP because the treaty does not provide for a balance of obligations and responsibilities between the nuclear weapon states and non-nuclear weapon states.

While the treaty had no provisions for imposing non-proliferation obligations on the nuclear-weapon states, it places full score safeguards and controls on the nuclear programmes of the non-nuclear weapon states.

NTP is discriminatory and legitimises the possession of nuclear weapons by nuclear weapon states.

NUCLEAR POWER

The total installed nuclear power capacity by the end of Eighth Five Year Plan is expected to grow 2,170.

Nuclear Power Share of Electricity Production 1986-89

	1990	1989	1988	1987	1986
France	74.5%	74.6	69.9	69.8	69.8
Belgium	60.1%	60.8	65.5	66.0	67.0
Hungary	51.4%	49.8	48.9	39.2	25.8
Republic of Korea	49.1%	50.2	46.9	53.3	43.6
Sweden	45.9%	45.1	46.9	45.3	50.3
Switzerland	42.6%	41.6	37.4	38.3	39.2
Spain	35.9%	38.4	36.1	31.2	29.4
Bulgaria	35.7%	32.9	35.6	28.6	30.0
Finland	37%	35.4	36.0	36.6	38.4
Germany, Fed. Rep.	33.1%	34.3	34.0	31.3	29.4
Czechoslovakia	28.4%	27.6	26.7	25.9	21.1
Japan	27.1%	27.8	23.4	29.1	24.7
United Kingdom	19.7%	21.7	19.3	17.5	18.4
United States	20.6%	19.1	19.5	17.7	16.6
Canada	14.8%	15.6	16.0	15.1	14.7
USSR	12.2%	12.3	12.6	11.2	10.1
Argentina	19.8%	11.4	11.2	13.4	12.2
German Dem. Rep.		10.9	9.9	9.7	9.7
South Africa	5.6%	7.4	7.3	4.5	6.8
Yugoslavia	5.3%	5.9	5.2	5.6	5.4
Netherlands	4.9%	5.4	5.3	5.2	6.2
India	**2.2**	**1.6**	**3.0**	**2.6**	**2.7**
Pakistan	1.1%	0.2	0.6	1.0	1.8
Brazil	1%	0.7	0.3	0.5	0.1
Italy		0.0	0.0	0.1	4.5

Expressed as percentage of total electricity produced.

Note:

(i) In Taiwan, China, the percentage for 1989 to 1986 are 35.2%, 41%, 48.5%, 43.8%

(ii) In 1990, five contries genrated cose to half or more of their total electricity using nuclear power. Altogether 12 countries relied upon nuclear power plants to supply at least one-forth of then home electricity needs.

NUCLEAR RESEARCH

A total of 323 research reactors are operating in 55 countries and another 13 are being built as of June 1990.

(Source: *IAEA Newsbrief*, 5(6) August-Sept, 1990)

Nuclear Power Status around the World Status as of 31 December 1990 with coresponding data for 1989 in parenthesis

	Operaational		Under construction	
	No. of units	Total net MWe	No. of units	Total net MWe
Argentina	2 (2)	935	1 (1)	692
Belgium	7 (7)	5 500		
Brazil	1 (1)	626	1 (1)	1 245
Bulgaria	5 (5)	2 585	2 (2)	1 906
Canada	20 (18)	13 993	2 (4)	1 762
China			2 (3)	2 148
Cuba			6 (2)	816
Czechoslovakia	8 (8)	3 264	(8)	3 336
Finland	4 (4)	2 310		
France	56 (55)	55 778	6 (9)	8 305
Germany	26 (30)	24 430	6 (6)	3 319
India	7 (7)	**1374**	7(7)	**1540**
Hungary	4 (4)	1 645		
Iran, Islamic Rep. of			2 (2)	2 392
Japan	41 (39)	30 917	10 (12)	9 012
Korea, Rep. of	9 (9)	7 220	2 (2)	1 900
Mexico	1 (1)	654	1 (1)	654
Netherlands	2 (2)	508		
Pakistan	1 (1)	125		
Romania			5 (5)	3 125
South Africa	2 (2)	1 842		
Spain	9 (10)	7 064		
Sweden	12 (12)	9 817		
Switzerland	5 (5)	2 952		
United Kingdom	37 (39)	11 506	1 (1)	1 188
USA	112 (110)	100 630	1 (4)	1 165
USSR	45 (46)	34 673	25 (26)	21 255
Yugoslavia	1 (1)	632		
World total*	423 (424)	325 873	83 (99)	65 760

* The total includes Taiwan, China, where six reactors totalling 4890 MWe are in operation. In 1990 they accounted for 35.2% of the total electricity generated there.

(Source: *IAEA Bulletin*, 1/1991)

O

OPERATION BLACK BOARD

Under the scheme to provide all primary schools in the country with essential physical facilities, a sum of Rs 147 crore has been allocated in 1990-91. It is expected that all primary schools in about 10 to 15 per cent blocks/municipal areas of the country would be covered.

An amount of Rs 372.98 crore was released by the Centre to States and Union Territories and Rs 581 crore was mobilised by States under the scheme of "Operation Blackboard", during the 7th Five Year Plan .

ORISSA GOVERNMENT

Government in Orissa has come under severe criticism from the Comptroller and Auditor General of India for making "inflated claims" on drought relief with Central assistance and for utilising huge amounts of it in non-affected areas in clear violation of Central guidelines.

State Government reported the expenditure for 1987-88 as Rs 52.86 crore against the Centrally fixed ceiling of Rs 59.61 crore, the actual expenditure as reported by units was only Rs 38.16 crore.

The Government was given extension up to June 1988 to make use of the unutilised assistance to the extent of Rs 6.29 crore on employment generation in the drought-hit areas, and on fodder cultivation. However, a substantial amount of the assistance remained unutilised.

Contrary to specific Central instructions that no contractor should be engaged for employment generation works, these were executed through contractors for irrigation and drinking water supply schemes.

Out of Rs 10.57 crore placed with the Orissa Lift Irrigation Corporation (OLIC) to take up modernisation of field channels and construction of lift irrigation project in districts, it spent only Rs 2.80 crore till the end of March 1988, which worked out to 27 per cent. While Rs 7.77 crore remained unutilised with the corporation, Rs 81.85 lakh of the Rs 2.80 crore remained unspent with the executing agencies.

The expenditure included Rs 31 lakh paid to the Orissa State Electricity Board for replacement of conductors which did not form a part of the employment generation component.

During 1987-88, all the 13 districts were declared drought-affected, and the districts of Bolangir, Kalahandi, Ganjam, Koraput and Phulbani were worst affected.

However, Cuttack district was given highest share of funds (Rs 2.15 crore), though it was not one of the worst affected districts.

Contrary to Central instructions that the feeding programme implemented under drought relief with Central Assistance should not cover Integrated Child Development Service (ICDS) blocks, huge amounts were spent on ICDS blocks in Cuttack and Koraput districts.

Besides, 15 blocks, not declared as drought-affected, were, covered under the programme.

The Centre approved Rs 43 crore for the employment generation scheme for 1987-88. However, the Orissa Government spent only Rs 27.02 crore but reported an expenditure of Rs 36.45 crore.

Orissa Government was required to send the list of works executed under the employment generation scheme to the ministries of Finance and Agriculture, but this was not done. Neither was any monitoring or evaluation study undertaken by the Government to assess the community assets created out of relief funds.

ORISSA LIFT IRRIGATION CORPORATION LIMITED (OLIC)

The Comptroller and Auditor General of India has severely indicted the State-owned OLIC for the wide discrepancies in its achievement claims and the reality. Though on paper, the achievement of the company's programme between 1982-83 and 1986-87 was of the order of 97 to 138 per cent (number of projects) and 76 to 112 per cent (irrigation potential), the projects were not available for utilisation to farmers.

P

PAKISTAN

Pakistani Nationals

As many as 7,677 Pakistani nationals were overstaying in various States as on December 1989.

The maximum number was from Maharashtra. The State authorities had been directed to take effective measures to locate and deport overstaying foreigners.

More than two lakh Pakistani nationals entered the country in the past one year with travel documents, while 9,768 stayed back after their visas expired.

Indian Defence Personnel in Pakistani Jails

Fifty-four missing Indian defence personnel are believed to be in Pakistani jails.

The question of locating and repatriating these missing Indian defence personnel had been taken up with Pakistan several times, including during the Foreign Secretary-level talks held in July and August, 1990.

The Government of Pakistan maintains that there are no missing defence personnel in their custody.

Pakistan Defence Expenditure

There has been an increase of 11.6 per cent in the defence budget of Pakistan for 1991-92, over that of last year.

PALMOLEIN

State Trading Corporation (STC) has already contracted to import about 66.000 tonnes of palmolein for the current financial year at a cost of Rs. 35 crore. In 1989-90, the STC imported 2.94 lakh tonnes of edible oils at a value of Rs 166 crore.

PARKING CHARGE PERMITS IN DELHI

The Delhi Administration's Directorate of Transport suffered a loss of Rs 2.38 lakh due to incorrect charging of permit fee for temporary permits and non-realisation of parking charges, according to the Comptroller and Auditor-General of India report for the year ended March 31, 1990.

During test-check of records of the Directorate of Transport by audit (May 1989), it was noticed that temporary permits for a period of 15 days or one month or two months were issued in 1,549 cases. Proportionate fee of Rs 25, Rs 50 and Rs 100 was charged for a period of 15 days, one month and two months, respectively at the rate of Rs 50 per month instead of Rs 5 per day for each permit.

Since the rules did not provide for charging of fee for temporary permits at monthly rate for periods of less than four months, the practice followed by the Directorate was irregular. The non-observance of rules correctly had resulted in loss of revenue amounting to Rs 1.62 lakh, the report pointed out.

The case was reported to the Home Ministry (May 1990), their reply has not been received till (December 1990).

The scooter/car stands at the offices of the Director of Transport, Rajpur Road and Tilak Marg, are being run by an Automobile Association from February 1983 on contract basis.

The parking charges are collected by the Association which are to be shared by the Director of Transport and the Association on 50:50 basis and the charges receivable by the lease for parking. A lest check of records in Audit revealed that the Directorate had not made effective efforts to recover its share of parking charges amounting to Rs 1.62 lakh due from the Association.

The Association deposited in May 1989 Rs 59,615 constituting 50 per cent Government share of parking charges for the period from February 1983 to March 1988 after erroneously deducting the expenses defrayed by it on the staff deployed on collection etc. Besides, annual dues for the period from April 1988 to March 1989 amounting to Rs 57,472 were not paid by the Association.

PAWAN HANS LIMITED

The Comptroller and Auditor General of India has criticised the under-utilisation as also grounding of helicopters by Pawan Hans Limited resulting in an estimated loss of nearly Rs 7.2 crore in 1986-87, Rs 7.44 crore in 1987-88 and Rs 7.89 crore in 1988-89.

There is no possibility of the company breaking even in the near future. Apart from six helicopters required for VIP communication squadron of the Air Force, the requirement of ONGC was initially assessed at 12 helicopters. ONGC requirement was subsequently increased to 21 helicopters in 1983. A fresh assessment of helicopter requirement was again made in 1985 when the demand on ONGC was increased to 25 and a demand of 17 helicopters for non-oil sector was also projected for the first time. The Ministry of Civil Aviation proposed in January 1986 to purchase of 21 Westland helicopters and 27 Dauphin helicopters including six helicopters for Air Headquarters, covering the requirements of oil and non-oil sectors.

Out of the 42 helicopters received by the company for its own operations, it could hardly deploy 25 helicopters as of March 1989. Taking into account maintenance reserve at 20 per cent of the regular fleet of 39 existing helicopters as on March 31, 1989 the company has at least six helicopters surplus to its requirement. This has resulted in extra expenditure to the extent of Rs 33.68 crore. In addition, the six helicopters costing Rs 25.77 crore, which were purchased for Air Headquarters Communication Squadron and which the Ministry of Defence ultimately declined to accept, are also lying idle from December 1987.

As against the average utilisation of 100 hours per month envisaged in the project report, the actual utilisation of Westland helicopters was around 40 hours and that of Dauphin helicopters 55 hours per month during the period from September, 1986 to March 1989. The company stated that the average utilisation of helicopters depended upon the requirement of customers. However, in terms of flying hours the company was able to fulfil the demand of ONGC to the extent of 86 per cent and 98 per cent in the case of Westland and Dauphin helicopters respectively. As regards the contractual demand of ONGC for Bombay offshore in terms of number of days for which helicopters were required to be provided by Pawan Hans, the level of satisfaction was 72 per cent in case of Westland and 90 per cent in case of Dauphin during 1988-89. This necessitated ONGC hiring helicopters from Indian Air Force ranging from 30 hours to 130 hours a month. It would thus be observed that although the company had surplus helicopters, it could not meet the demand of ONGC fully at Bombay.

Westland helicopters experienced 342 reportable defects and six Westland helicopters were grounded for 20 times each for such defects between December 1986 to March 1989; there were 46 reportable defects in Dauphin helicopters over the same period of time. The incidence of unscheduled repair works out to one after 51 hours of flying in the case of Westland and 619 hours of flying in case of Dauphin Manufacturers took unduly long time in repairing engines which were with drawn prematurely. Ten engines in case of Westland and three engines in case of Dauphin took more than six months for repair.

While the cost of maintenance and repair per flying hour was initially estimated in February 1986 at Rs 5,901 for Westland and Rs 3,981 for Dauphin, it subsequently increased considerably and was estimated by the manufacturers in February 1989 to be Rs 21,588 for Westland and Rs 9,375 for Dauphin. The company has lost three helicopters due to accidents up to March 1989. Two of the helicopters were Westland and one Dauphin.

The company is constrained by a severe shortage of pilots and engineers. As against company's requirements of 140 pilots and 73 engineers, they had only 92 pilots and 20 engineers in position as on March 1989. Due to acute shortage of engineers the company is forced to obtain temporary exemption from Director General of Civil Aviation for allowing its technicians to sign flight release certificates.

PENSIONERS

Neutralisation of the price rise is 100 per cent to pensioners drawing pension up to Rs. 1750 and 75 per cent to those in the range of Rs. 1751 to Rs. 3000 and 65 per cent to those who are drawing pension over Rs. 3000.

PETROLEUM

Consumption

The growth in consumption of petroleum products has decreased from 8.1 per cent in 1989-90 to 1.3 per cent in 1990-91 due to a number of factors.

Conservation

The Government has decided to intensify efforts at conserving petroleum products and their proper utilisation in the country in view of the fact that the share of oil and natural gas as commercial energy has been increasing, and their growing demand has to be met by importing these at the cost of foreign exchange.

Since the recent increase in the prices of petrol and diesel has not appreciably curbed the demand for petroleum products, officials feel that intensification of efforts at conserving these has become more imperative than ever before.

A special cell has been constituted for coordinated and effective functioning of all the agencies engaged in conserving petroleum products.

The Joint Secretary in charge of Conservation of Petroleum Products has been made Vice-Chairman of the Petroleum Conservation Research Association, as also Chairman of its executive committee.

Till the end of the 1989-90 financial year, studies carried out in small industrial units, said 750 kls of petroleum products valued at Rs. 25.50 lakh were expected to be saved, the report said.

Up to December 1989, 717 follow-up studies in respect of fuel consumed by big industries consuming 250 kls or more fuel annually were conducted. They revealed additional savings achieved on account of implementation in 1989-90 of the recommendations in the previous year was Rs. 14.62 crore. By end of March 1990, petroleum products worth Rs. 18 crore were expected to be saved, the report said.

In the agricultural sector, the PCRA continued the programme of rectification of lift irrigation pump sets in Gujarat, Rajasthan, Uttar Pradesh, Punjab and Andhra Pradesh with the help of the agencies concerned.

A total of 194 pump sets have been rectified which save petroleum products worth Rs. 0.14 lakh till December 1989. An estimated six hundred pump sets had been rectified at the end of the financial year.

As an important part of research and development, the PCRA has designed a fuel-efficient melting furnace which has been installed in a manufacturing unit at Shahdara in Delhi. The project was expected to result in fuel saving of about 15 per cent.

Efforts have also been initiated to explore the possibility of conserving petroleum products in the sectors not conserved by the PCRA so far. A working group has been constituted to study and make recommendations on conserving petroleum products in the shipping sector. An examination of the possibility of effecting economy in the ocean transportation of petroleum products has also been taken up to identify specific areas for taking action.

(Source: *UNI Release)*

Prices

The whopping hike in petrol prices by 25 per cent from Rs. 9.84 to Rs. 12.23 per litre exposed the Government's failure to contain the priceline. This is all the more disturbing as the price hike comes within only months of the last price rise when per litre petrol cost was moved up from Rs. 8.50 to Rs. 84. The new hike will further escalate the already rapidly rising rates of all essential commodities.

Three-wheeler and taxi drivers have started charging higher fares ranging from 20 to 30 per cent over the earlier rates. With no proper price rating announced by the authorities, the people have to bear the brunt of arbitrarily demanded fares by them.

Petrol price, which was Rs. 1.50, rose

to Rs 9.80 on March 20, 1990 and again to Rs 12.23 on October, 1990.

PILOTS

There are 348 pilots in Air India of which nine belong to the Scheduled Caste category. In Indian Airlines. there are 516 pilots of which 16 belong to the Scheduled Caste and three to Scheduled Tribes.

The Indira Gandhi Rastriya Uran Akademi will offer 100 per cent scholarship of the training fee to three Scheduled Caste and Scheduled Tribes candidates from the seventh CPL course onwards and would also recommend to the concerned State Government for grant of scholarships to meet the expenses on board, lodging, library and sport of SC and ST candidates belonging to the State.

POLICE

Expenditure on Advertisement

Delhi Police had spent Rs. 1,12,46,184 on advertisements through newspapers and Doordarshan during the 1990. About 10,000 door chains and 12,000 magic eyes had been fixed in various areas during the crime prevention weeks. These were provided free of cost by various organisations.

Corruption and Misbehaviour

As many as 13 Delhi Police officials were dismissed between 1989 and 1990, while the services of 12 were forfeited, 97 were censured and 50 warned after departmental inquiries were initiated against 135 officials during this period 16,8000 complaints were received in 1988 against Delhi Police personnel for misbehaviour and misuse of their authority, 17,143 in 1989 and 20 179 in 1990.

Criminal Personnel

A constable (driver) and his father, a sub-inspector, were found involved in the bomb explosions in Rajouri Garden and Tilak Nagar police stations on April 10, 1990. The sub-inspector has been arrested and his son and declared a proclaimed offender.

POLLUTION

Pollution in Arbian Sea

The research ship "Sagar Sampada" which was sent to the Arabian Sea did not find any evidence of pollution resulting from the Gulf war.

Saikia Panel

A panel on vehicular pollution control has been set up under the chairmanship of Justice S.N. Saikia to make an assessment of the technologies available for vehicular pollution control in the world and also in India for controlling vehicular pollution.

Pollution in Delhi

Delhi was one of the highest polluted cities around the world with its pollution rating 871.92 tonnes at present. Delhi Transport Corporation buses contributed 55 to 60 per cent of the air pollution.

The DTC had prepared a comprehensive action plan to help reduce the air pollution. It was embarking on a phased programme of bus fleet renewal to withdraw the order buses from service. It also intended to instal smoke meters to check exhaust pollution from its vehicles at the maintenance bases.

The fitness certificate of 3,250 vehicles had been cancelled and about 2,000 other vehicle owners prosecuted in the drive against exhaust pollution in Delhi. About one lakh vehicles had been checked for pollution.

The Central Pollution Board, monitoring the ambience levels of pollution in the Capital, had found it ranging as high as 1,500 mg at times against a permissible level of 1,000 mg.

The Government has entered into an agreement with the World Bank to provide $ 147.4 million financial assistance for an industrial pollution control project.

PORTS

The total approved outlay for major ports for 1990-91 is Rs. 472.04 crore of which Rs. 372.28 crore are for continuing schemes and Rs. 99.66 crore for new schemes for improving the existing port infrastructure and expansion of facilities.

Free Port

The Government has set up an advisory committee to examine the desirability and feasibility of establishing a free port in India. A sub-committee thereof had preliminary discussion with the Tamil Nadu Government when some alternative sites were considered.

POST OFFICES

There were 1,47,102 post offices in the country as on December 31, 1989, with 1,30,838 in rural areas and 16,264 in the urban areas.

Kerala has the highest density of post offices among all States in the country. There are more than 12 offices in the State for every 100 square km as against the national average of 4.5 post offices.

POSTAL RATES (REVISED)

The revised rates of various postal services, as announced by the then Finance Minister Madhu Dandavate while presenting the 1990-91 budget in March 1990, became effective from June 11, 1990.

The existing and revised rates of some of important postal services are:

	Existing Rate	**Revised Rate**
Inland Letter Cards:	35 paise + 15 paise stationery charge.	75 paise. (No Stationery Charge)
Letters (Envelopes):	60 paise for first 10 gms and 40 paise for every addl. 10 gms 15 paise sty. charge for embossed envelope.	Re 1 for every 20 gms weight or part thereof. No sty. charge.
Book Post: (i) Book pattern and sample packets (ii) Book packets containing printed books	 50 paise for first 60 gms and 50 paise for every addl 50 gms. 30 paise for first 100 gms + 30 paise for every addl 100 gms.	 Re 1 for first 50 gms and Re 1 for every addl 100 gms. 50 paise for first 100 gms + 50 paise for every addl 100 gms
Parcels:	Rs. 4 for every 500 gms.	Rs. 6 for every 500 gms.
Registration:	Rs. 5	Rs. 6
Acknowledgement due:	50 paise	Re 1
Money orders:	For M.Os upto Rs 50 40 paise for every Rs. 10 or fraction thereof subject to a minimum of 50 paise. For M.Os above Rs. 50 50 paise for every Rs. 10 or part thereof.	Re 1 for every Rs. 20 or part thereof.
Certificate of Posting:	20 paise for 3 articles or less	Rs. 2 for 3 articles or less.
Post Box or Bags: (a) If taken separately. (b) If taken together.	 Rs. 60 per annum. Rs. 100 per annum.	Rs. 150 per annum. Rs. 250 per annum.

POVERTY

Poverty Line

An estimated 29.2 per cent of India's population lived below the poverty line in 1987-88. The percentage of population living below the poverty line had been declining during the last decade. In 1987-88 the percentage was estimated to be 48.3.

As many as 237.67 million people were estimated to be living below poverty line in 1987-88.

The surveys carried out by National Sample Survey indicate that the percentage of persons below poverty line has come down from 72.43 in 1977-78 to 58.40 in 1983-84 in rural areas and from 52.55 in 1977-78 to 39.90 in 1983-84 in urban areas.

Rural Poverty

The Planning Commission has estimated that 48.3 per cent of the rural population in Orissa was below the poverty line as compared to 33.4 per cent in the entire country. The level of rural poverty in Bihar was 42.7 per cent.

Regarding the level of other Indicators of backwardness of rural areas, Bihar was backward but not the worst affected. Bihar's level of rural literacy was better than that of Jammu and Kashmir, Madhya Pradesh and Rajasthan, the Minister said.

Bihar had a lower level of infant mortality than Gujarat, Madhya Pradesh, Orissa, Rajasthan and Uttar Pradesh.

Urban Poverty

Alleviation of urban poverty was given a boost during 1990-91 through a package of programmes such as Nehru Rozgar Yojana, Urban Basic Services for the Poor and Environmental Improvement of Urban Slums.

The Nehru Rozgar Yojana was launched in October 1989 and was recast with new guidelines the following year. Its three components are urban micro enterprises, urban wage employment and the scheme of housing shelter upgradation.

The scheme of urban micro enterprises assists the poor in upgrading their skills and to set up self-employment ventures. An amount of Rs. 432.61 lakh was sanctioned as loan and subsidy for setting-up micro enterprises upto Dec. 31, 1990. More than 15,000 beneficiaries have been trained or are undergoing training in various trades.

The urban wage employment scheme provides wage opportunities to the urban poor by utilising their labour for construction of socially and economically useful, public assets in the jurisdiction of urban local bodies. By Dec. 31, 1990, an expenditure of Rs. 1445.84 lakh has been incurred on public works in various States and Union Territories and, they had generated more than 31 lakh mandays of employment.

The scheme of housing and shelter upgradation seeks to provide assistance for housing and shelter to the economically weaker sections of the urban population as well as to provide opportunities for wage employment and upgradation of construction skills. The HUDCO has sanctioned 170 schemes for housing and shelter upgradation under the scheme upto the end of 1990 and they are worth Rs. 80.88 crore involving construction or renovation of 2.60 lakh dwelling units.

Another programme for the poor is the urban basic services programme and it is fully financed by the Central Government and an outlay of Rs. 23 crore has been earmarked for it. The basic services are being provided to 169 towns and it is proposed to extend the same to 247 new towns.

The scheme of providing shelter to the pavement dwellers has been extended to urban areas wherever the problem exists. During the fist nine months of 1990-91, as many as six schemes were sanctioned by the HUDCO for construction of night shelters benefiting 4640 footpath dwellers. Another 22 schemes for the benefit of 2868 pavement dwellers are under scrutiny and likely to be sanctioned soon.

(Source: *Annual Report*, Ministry of Urban Development, New Delhi)

POWER

There was a big shortfall in the achievement of the target for additional capacity creation in the power sector in 1990-91.

As against a target of 4,212 MW for 1990-91, only 2,776.5 MW of additions were made to generating capacities by the Central and State sector agencies taken together.

The shortfall, which was around 35 per cent was on account of a failure in all the three hydro modes of generating power electric, thermal and nuclear to hit the targets fixed for them.

Funding had proved to be one of the biggest hurdles in the way of additional capacity creating.

As against a programme of commissioning 70 units with capacity aggregating 4,212 MW in 1990-91, only around 31 units could be commissioned during the year in the Central and the State sectors taken together.

In the current year, the priority had been altered "a little" to give greater stress on renovation and modernisation (R and M) of power stations.

While in the earlier scheme of things, creation of fresh capacity was accorded a higher priority, now the emphasis was on optimal utilisation of existing capacity.

Equipment problems had also come in the way of the capacity creation target being fulfilled.

For the current year, a fresh capacity creation target of 3,875 MW has been fixed of which 700 would be hydel power.

The target 1990-91 was 1006.5 MW for hydel capacity in the State sector, 2,970.5 MW for thermal capacity (of which 732 in Central and rest under State sector) while 235 MW was the target for nuclear capacity creation.

As against this programme, the achievement was 445.5 for hydel, 2331 MW in thermal (the Central sector hit its target) while the nuclear capacity was missed altogether.

Among the units which were commissioned in 1990-91 were the 115 MW capacity Varahi hydel project in Karnataka, the 2·150 MW Bansagar hydel units in Madhya Pradesh and the 60 MW Kadana hydel project in Gujarat.

In the thermal sector, the 210 MW Bokaro 'B' unit of Damodar Valley Corporation was commissioned, alongwith the 210 MW Vindhyachal project of the National Thermal Power Corporation in Madhya Pradesh and the fourth 210 MW of the Neyveli Lignite Corporation in Tamil Nadu.

Altogether 18 projects in the thermal power generation in the State sector were commissioned during the year under review with capacities varying from a single megawatt (diesel generating sets at Manipur) to 500 MW.

(Source: *PTI Release)*

Hydel Power

The total hydro-electric potential of the country is estimated at 84,077 MW (at 60 per cent load factor). Out of this, 11,875.72 MW (at 60 per cent load factor) was developed and 5,942 MW was under various stages of development.

Schemes totalling 2,686,35 MW had been cleared by the Central Electricity Authority.

Generation

Power generation in the country had gone up from 2.21 lakh MW in 1988-89 to 2.64 lakh MW in 1990-91.

Gap

The availability of electricity in the country during April to June 1919 was 62,772 million units as against the requirement of 68,815 million units, resulting in a shortage of 8.8 per cent.

Power Theft in Delhi

The 494 cases of power theft detected during January to June 1991, 189 cases pertaining to East Delhi. HE said 160 FIRs had been lodged with the police, of which 55 pertained to East Delhi.

PRAWN EXPORTS

The country earned Rs. 470.33 crore by way of prawn exports in 1988-89, against Rs. 425.78 crore in the previous year.

PRESIDENT'S EMOLUMENTS

Parliament on June 1, 1990, approved the President's Emoluments and Pension (Amendment) Bill seeking to raise the emoluments of the President to Rs. 20,000 per month and the pension of ex-President to Rs. 10,000 per months. The Bill also seek to allow former Presidents facilities like free accommodation with free electricity and water, telephone and care facilities and secretarial assistance on an enhanced scale.

PRICE RISE

See also Inflation.

Prices increased sharply, rising by as much as 9.7 per cent during the tenure of the National Front Government, an analysis of official figures show.

The variation in prices as captured in the movement of the official wholesale price index (base: 1981-82-100) reveal that in the period between December 4, 1989 and October 27, 1990 (latest available) the rise was particularly sharp for primary articles—higher by 13.2 per cent.

Under primary articles, prices of food articles witnessed an even sharper rise at 14.2 per cent. Prices of cereals rose by 6.7 per cent, of pulses by 9.4 per cent, of vegetables by 33.8 per cent and of fruits by 28.3 per cent.

The figures released by the office of the Economic Advisor, Ministry of Industry, show that the rise has been particularly sharp during the current year. On an average weekly basis between April and October this year, while the overall price level rose by 7.4 per cent, prices of primary articles were higher by 12 per cent and of food articles by 15.5 per cent.

Over the same period in the previous year, the overall rise was 7.5 per cent, while the prices of primary articles and food articles were higher by 6.4 per cent and 5.2 per cent, respectively.

Further analysis reveals that in the first six months of the year ended September 1990, prices of food articles continued to rise, despite munificent monsoons and record foodgrains production of 170.3 million tonnes in 1988-89 and 171.2 million tonnes in 1989-90.

Under food articles, prices of foodgrains rose by a massive 10 per cent, as against 3.8 per cent over the same period last year. Significantly, prices of cereals rose by 10.2 per cent, as against 1.5 per cent last year, while those of pulses witnessed a deceleration, rising by 8.8 per cent compared to 15.1 per cent in the previous year.

The pervasive trends in the prices of primary food articles would be particularly severe on those sections of the population, where wages are not indexed to inflation, namely, those depending on wage labour for livelihood.

In the first six months of the current year (April to September), there was a rise in the prices of rice by 9.1 per cent, as against 7.8 per cent in the previous year. Wheat prices increased by a hefty 15.8 per cent compared to a decline of 8.1 per cent in the previous year.

Prices of coarse cereals witnessed mixed trends. Those of jowar declined by 8 per cent, as against a rise of 4.2 per cent in the previous year, while those of bajra were up by 20 per cent, as against 1.8 per cent, of maize by 5.9 per cent as against a decline of 5 per cent.

Over the same period, under pulses, prices of gram were up by 13.8 per cent, of arhar by 17.1 per cent, of masur by 7.2 per cent, and of urad by 2.9 per cent, while those of moong declined by 6 per cent.

Interestingly, prices of manufactured products, the largest group in the index, witnessed relatively subdued trends, with the exception of edible oils. As a result, it has to an extent cushioned the impact of the steep rise in prices of primary articles witnessed in the first six months ended September 1990.

Prices of manufactured food products more by 8.3 per cent compared to a massive 19.6 per cent in the previous year. Significantly, sugar prices declined by 0.5 per cent as against 19.4 per cent in the previous year.

Edible oil prices on the contrary rose steeply by 15.6 per cent between April and September this year, as against 12.7 per cent in the previous year.

This has been largely attributed to expectations of lower production this year. This coupled with the fact that the Government would be unable to make any large scale imports in the light of the balance of payments crisis has affected market sentiments.

Among other manufactured products, prices of intermediates, namely steel and aluminium, witnessed a perceptible slowdown in the first six months ended September.

In this period, prices of basic metals and alloys rose by 3.7 per cent as against 8.3 per cent in the previous year, while non-ferrous metals rose by 4.8 per cent and 5.1 per cent, respectively.

There has been increase in the prices of essential commodities. The wholesale price index has gone up from 167.5 in November 1989 to 184.3 in November 1990. The consumer price index has gone up from 176 in November 1989 to 195 in October 1990.

PRIVATE SHIPS

Between July 1, 1988 to end-July 1989, private sector shipping companies had acquired/ordered 45 ships of total value of Rs. 592 crore, while the Shipping Corporation of India had 10 vessels of a total value of Rs. 362 crore during this period.

PROCUREMENT OF RICE AND WHEAT

Rice procurement during the current kharif marketing season of 1989-90 has achieved the all-time record.

As on May 25, 1990, total quantity of 111.43 lakh tonnes of rice, including paddy in terms of rice, had been procured as against 73.62 lakh tonnes during the corresponding period in 1988-89 and the earlier record procurement of 98.63 lakh tonnes of rice in 1984-85 and 98.76 lakh tonnes in 1985-86. During the current kharif marketing season endings on September 30, 1990 end the total procurement of rice was expected to be around 112 lakh tonnes.

The procurement of wheat during 1990-91 rabi marketing season which commenced on April 1, 1990 has already crossed the last year's level of 90 lakh tonnes. As on May 25, 1990, a quantity of 97.14 lakh tonnes of wheat had been procured against 80.07 lakh tonnes during the corresponding period in 1989-90. The total procurement is likely to exceed 100 lakh tonnes.

The record procurement of rice and wheat during the current kharif and rabi marketing season will enable the Government to meet the demands of public distribution system as well as to replenish the buffer stocks of rice and wheat.

The enhanced procurement prices announced by Government have contributed to the increase in procurement of rice and wheat.

(Source: *Hindustan Times)*

PRODUCTION INDEX

Industrial

According to the quick estimates of Index of Industrial Production (base 1980-81=100) issued by the Central Statistical Organisation, the general index for the month of January 1990 stands at 215.3, which is higher by 11.7 per cent as compared to the corresponding month of 1989.

Taking into account the IIP figures released up to December, 1989, and the quick estimate of the index for January, 1990, the average index for the first 10 months of the financial year 1989-90 works out to 1888.8 which is higher by 6.3 per cent as compared to the index of 177.6 for the corresponding period last year.

(Source: *Hindustan Times*)

PROVIDENT FUND

About 99 crore was in arrears towards employees provident fund contribution from the unexempted establishments as on March 31, 1990.

PUBLIC DISTRIBUTION SYSTEM

See also Fair Price

PUBLIC SECTOR ENTERPRISES

Chairmen

The post of Chairmen and Managing Directors in 29 central public sector undertakings remained vacant at present.

These include the National Textile Corporation, Air India, Vayudoot, Engineers India Ltd. Neyveli Lignite Corporation, Bhilai Steel Plant, India Tourism Development Corporation, Modern Food Industries, and Fertilisers and Chemicals Travancore Ltd (FACT).

Memoranda of Understanding (MOU)

11 public sector enterprises signed

memoranda of understanding (MOU) with the Government and the net profit of the 11 MOU companies increased to Rs. 24800.61 crore from Rs. 1991.78 crore in 1987-88.

Profits

One hundred and four operating Central public enterprises out of 186 have earned a net profit of Rs. 5,100 crore during the financial year 1989-90. During 1988-89 only 101 units earned net profit of Rs. 4,734 crore.

One hundred eighty six operating Public Sector enterprises have earned a net profit of Rs. 3,427 crore during the financial year 1989-90 as against a net profit of Rs. 2,867 crore earned by them during the preceding year 1988-89, showing an increase in net profits of Rs. 560 crore (19.54 per cent).

It is expected that in the final tally the overall net profits in Central PSEs would be around Rs. 3,500 crore against Rs. 2,981 crore during the corresponding period of the previous year. Eighty two enterprises incurred a net loss of Rs. 1,674 crore during 1989-90 which is less by Rs. 194 crore compared with the net loss of Rs. 1,867 crore incurred by enterprises during 1988-89.

During the year non-oil sector earned a net profit of Rs. 751 crore against Rs. 307 crore in the previous year showing improvement in profits by 144.6 per cent. It is mainly because of NMDC and NALCO in the Minerals and Metals Sector, NTPC in the Power Sector, SCI in the Transportation Services Sector and Power Finance Corporation in the Financial Services Sector each of these improved their net profit by over Rs. 50 crore as compared to last year. In the oil sector, profits in 1989-90 were Rs 2,676 crore as against Rs. 2,560 crore in 1988-89 showing an increase of 4.5 per cent.

Fourteen out of 20 non-oil sectors have shown improvement in their profitability by either increasing the profits or reducing the losses during the year.

(Source: *Hindustan Times*)

The profits of Central public sector undertakings for the year 1990-91 shows a sharp decline of about 38 per cent when compared to the previous year.

189 out of PSUs about whom provisional data has been received earned a net profit of Rs. 2,730.27 crore during 1990-91 against Rs. 3,748 crore earned during the previous year. This means a decline in net profits of Rs. 1,017,73 crore (37.28 per cent). The targeted net profit loss during 1990-91 for these enterprises was Rs. 3,115.09 crore.

The significant variation from the target occurred in the petroleum, coal and lignite, fertiliser and heavy engineering sectors. Among the 189 enterprises, 105 earned a net profit of Rs 4,916.60 crore during 1990-91 against a target of Rs. 5,018.64 crore which is lower by Rs. 671.35 crore when compared to the net profit of Rs. 5,589.95 crore earned by these enterprise during the previous year.

84 PSUs incurred a net loss of Rs. 2,188.33 crore against a target of Rs. 1,903.55 crore which is higher by Rs. 346.38 crore when compared to the net loss of Rs. 1,841.95 crore incurred by them during 1989-90. The net profit as a percentage of capital employed declined to 2.66 in 1990-91 from 4.62 in 1989-90.

Gross sales at current prices increased from Rs. 100,407 crore in 1989-90 to Rs. 111,587.33 crore in 1990-91 showing an increase of 11 per cent. However, gross sales as a percentage of capital employed declined to 108.52 in 1990-91 from 123.87 in the previous year.

An enterprisewise analysis of the financial performance indicates that 12 enterprises have shown an increase in profitability or decrease in losses by Rs. 20 crore or more during the year. These include NTPC, Power Finance Corporation, MMTC, Pawan Hans, Cement Corporation of India, Engineering Projects India and Indian Oil Corporation.

On the other hand, 18 other enterprises including ONGC, Indian Airlines, Hindustan Fertiliser Corporation, Oil India, Coal India Limited, Mahanagar Telephone Nigam Limited, Delhi Transport Corporation and BHEL have shown a decline in their profitability or increase in losses by more than Rs. 20 crore.

Sectorwise analysis indicates that petroleum leads among the profit-making sectors with a net profit of Rs. 2,275.03 crore during 1990-91 as compared with Rs. 2,896.33 crore earned during the previous year. It is followed by power (Rs. 666.01 crore), financial services (Rs. 281.14 crore), minerals and metals (Rs. 196.98 crore), telecommunication services (Rs. 179.43 crore) and trading and marketing (Rs. 164.55 crore).

Non-oil sector earned a net profit of Rs. 455.24 crore during 1990-91 as against Rs. 863.37 crore earned during the previous year thus showing a decline in profit by Rs. 410.03 crore (47 per cent). In the oil sector, the profit in 1990-91 was Rs. 2,275.03 crore as against Rs. 2,896.33 crore (21.5 per cent).

Ten out of 20 non-oil sectors have shown an improvement in their profitability by either increasing the profits or reducing the losses during the year. Financial services lead among these sectors with an increase in profit of Rs. 153.34 crore followed by power (Rs. 72.83 crore).

Among the loss-making sectors, the fertiliser sector continues to head the list with Rs. 355.94 crore followed by consumer goods (Rs. 211.74 crore), textiles (Rs. 193.61 crore) and contracts and construction (Rs. 116.12 crore). The decline in net profits during 1990-91 is mainly attributable to payment of arrears arising out of the wage revision during the year amounting to Rs. 1,000 crore, the loss due to the sharp upward revision in the exchange rate of foreign currencies amounting to Rs. 600 crore as well as the Gulf crisis.

Forty of the forty-five taken over enterprises from whom data has been received have incurred a net loss of Rs. 446.16 crore during 1990-91 as compared to a net loss of Rs. 475.16 crore incurred during 1989-90.

(Source: *Annual Report*, Department of Public Enterprises, New Delhi)

PULSES

The area under pulses cultivation has not declined in the country in the past two decades. It has fluctuated between 20.92 million hectares to 24.45 million hectares. The production of pulses in 1950-51 was 8.4 million tonnes, which rose to 13.85 million tonnes. The per capita availability of pulses was 40.4 grams per day, against the requirement of 47 grams per day.

PUNJAB KILLINGS

A total of 939 people, including 128 security personnel, were killed in Punjab till June 1990.

The number of killings during the period in 1989 and 1988 has been 564 and 1,266 respectively.

Militant Killings and Police Killings

(September 1981 to December, 1989.)

Political regime	Militant			Police		
	Militant killings	Monthly average	Per cent increase/ decrease	Police killings	Monthly average	Per cent increase/ decrease
Phase-1	40	1.54		26	1.00	
Phase-2	949	41.26	2581.96	870	37.83	3682.61
Phase-3	808	42.53	3.07	197	10.37	-72.59
Phase-4	3945	127.26	199.25	1809	58.35	462.81
Phase-5	139	139.00	9.23	28	28.00	-52.02
	5881	58.81		2930	29.3	

Notes:

Phase-1 Darbara Singh September 1981 to October 1983 (26 months).
Phase-2 Governor's Rule November 1983 to September 1985 (23 months).
Phase-3 S.S. Barnala October 1985 to April, 1987 (19 months).
Phase-4 S.S. Ray May 1987 to November 1989 (31 months).
Phase-5 N.K. Mukerji December 1989 (1 month).

Civilians, Militants and Security Men Killings

(January 1, 1990-November 11, 1990)

Total civilians killed	1770
Militants killed	1168
Security men killed	452
Total	3390

Note: This period include first five months of Government Rule under Phase-5.

6,280 persons were killed in Punjab since 1981 till end July 1990 in the continuing violence in the State. The highest number of killings took place 1988 when 1,949 person lost their lives.

The number of civilians, securitymen and terrorists killed in Punjab between June and November 1990 far exceeded the killings in the preceding six months. While 752 civilians and 133 security men were killed by terrorists during the six months between December 1989 and May 1990, the number increased to 1198 and 297 respectively during the period between June and November 1990. The number of terrorists killed also rose from 391 to 828 during this period.

As many as 7,600 people were killed in terrorist violence in Punjab in 1989, 1990 and up to May this year

While 4,742 people were killed by terrorists, 2,858 terrorists were killed. The respective break-up year-wise is: 1989—1,168 and 699; 1990—2,467 and 1,321; up to May 1991—1,107 and 838.

A total of 517 persons wee killed by the terrorists in Punjab in May and June this year.

23 candidates for the Assembly and the Lok Sabha were killed in the State up to July 4.

During the period 327 terrorists were killed in encounters and 349 arrested.

See also Terrorists.

R

RAIDS

As many as 1,057 searches were conducted in the first four months of 1990-91 and the value of assets seized was Rs. 33.23 crore and concealed income disclosed was Rs. 40.17 crore. In 1989-90, as many as 3,984 searches were conducted and value of assets seized was Rs. 128.02 crore and the concealed income disclosed amounted to Rs. 193.44 crores.

See also Corruption.

RAILWAYS

Computerisation

The Indian Railways has undertaken expansion of computerisation of passenger reservation system at Pune, Patna, Jammu and Tawi, Gorakhpur, Trivandrum, Bhubaneshwar, Cuttack and Jaipur, during 1990-91.

Contract with Malaysian Railways

The Indian Railway Construction Company Limited (IRCON) has signed a contract with the Malaysian Railway Administration on August 10, 1990 for the doubling of 88.3 km of railway track between Rawang and Seremban in Malaysia.

Derailments

See under Derailments.

Electrification

The Railways electrified 902 route kilometres (RKMs) during 1990-91, bringing the total electrification to 10,154 km.

Railway Freight

The Railways carried an all time record revenue earning freight traffic during 1990-91. The total revenue earning freight traffic lifted was 318.02 million tonnes, as against 309.97 million tonnes of such traffic during the previous year, thus registering an increase of 8.05 million tonnes. This exceeded the revised target of 316 million tonnes of revenue earning freight traffic. Movement of coal was 135.06 million tonnes, raw material for steel plants 25.90 million tonnes, pig iron and finished steel from steel plants 10.01 million tonnes, iron ore for export 13.14 million tonnes, cement 28.76 million tonnes, foodgrains, 25.29 million tonnes, fertilizers 18.31 million tonnes, petroleum and oil 25.05 million tones and other goods 36.50 million tonnes.

(Source: *Yojana*, July 15, 1991)

RATION CARDS IN DELHI

2,17,915 ration cards have been issued by the Delhi Administration to the jhuggi dwellers till April 17, 1990.

7,479 food cards involving 61,114 cereal and 36,230 sugar units were cancelled in Delhi during the last three years as the card holders were not found residing at the indicated address at a subsequent stage.

READERSHIP OF NEWSPAPERS

Altogether 13 per cent adults in Eastern India read newspapers or magazines, making it the country's lowest readership ares, according to a recent finding of the Operation Research Group (ORG).

The readership in South India is the highest (30 per cent), followed by West (23 per cent) and North (19 per cent).

Within the East Zone, West Bengal has the highest readership (20 per cent) followed by Orissa (11 per cent), Bihar (9 per cent) and Assam and other North-Eastern States (9 per cent).

(Source: *Hindu)*

Only 18 per cent of adults in Andhra Pradesh read newspapers or magazines and about 55 per cent of urban adults are exposed to the print media as compared with only seven per cent in the rural areas.

Variation in readership also existed among males (24 per cent) and females (12 per cent).

The readership of newspapers in the State was quite low as compared with Kerala (59 per cent), Tamil Nadu (31 per cent) and Karnataka (24 per cent).

Of the 71 lakh adults who read newspapers and magazines in the State, 91 per cent read Telugu publications, about 27 per cent read English and seven per cent read Urdu.

The readership of Tamil, Malayalam and Kannada publications was, however, low in the State, the study reveals.

Two-thirds of Telugu publication readers and 91 per cent of English readers lived in urban areas.

About 66 per cent of readers in Andhra Pradesh read dailies. The readership of weeklies, fortnightlies and monthlies was recorded as 61.11 and 41 per cent respectively.

((Source: *PTI Release)*

RESERVATION FOR SC/ST

The backlog of reservations for Scheduled Castes and Scheduled Tribes in the Central Government services as reported by various Central Government Ministries and departments in the last three years was 111,079.

To clear the backlog of Scheduled Castes and Scheduled Tribes vacancies, executive instructions were issued in April, 1989 banning dereservation in direct recruitment. These instructions were considered adequate.

See also Pilots.

RESERVE BANK OF INDIA

Growing fiscal deficits in the March budget touched a record of 2.4 per cent of gross domestic product (GDP). Coupled with heavy internal borrowings — the Government is mostly being run with borrowed money — its long-term external debt has been mounting rather alarmingly, standing at Rs. 81,168 crore on March 31, 1990 as against Rs. 69,681 crore in March 1989. While noting satisfactory economic growth (real GDP of 4.5 per cent) as "indeed creditable", the RBI is hopeful that the economy will be able to achieve a 5 per cent GDP growth in the current year in view of the bright outlook in foodgrains production. In what is perhaps a pointed reference to the ministerial predominance in monitoring public sector banking, a role that is traditionally supposed to be discharged by the Reserve Bank, minimising the "existing overlap of functions and responsibilities" between the Government and the RBI to help the latter to discharge its role more effectively as a regulator of the financial system. It is talks of greater autonomy for central banks in the context of a worldwide debate on monetary management. "Autonomy in the Indian context", the report goes on to say, "would mean that the Reserve Bank should be able to effectively discharge its role as a monetary authority and as a regulatory body. . . ."

The slide-down in the public sector banking management has a lot to do with the RBI being made to play second fiddle to the banking department of the Finance Ministry. The report has sought a fundamental change in the system of Government financing and any reforms that may be contemplated could be undertaken within the current framework of the RBI Act.

(Source: *Annual Report*, Reserve Bank of India, quoted in *Hindustan Times*)

RBI Retirement Age

The Reserve Bank of India has decided to increase the age of retirement of a section of its officers from 58 year to 60 "on a selective basis".

The decision will apply to officers recruited or promoted before July 19, 1969, and those workmen recruited earlier but promoted after the date. The new policy was in line with the practice prevalent in "a majority of public sector banks".

REVENUE

Revenue Collection

The figures collected at the end of the first four months of 1990-91 shows that revenue collection at the end of July 1990 has increased compared to the corresponding period in the previous financial year.

Income-tax collections of Rs. 818.89 crore represent an increase of 12.13 per cent over the collections during the corresponding period of last year. During the month, the income-tax department conducted 305 searches and the seized valuable assets were Rs. 1,062.50 lakh.

The collection of customs and excise duty totalled Rs. 13,344.24 crore which shows an increase of 12.87 per cent over the collections during the corresponding period last year. During the month, the customs department seized assets worth Rs. 2,863,60 lakh including gold, diamond and other valuable items.

(Source: *Hindustan Times*)

Central Revenue Expenditure (Percentage Distribution)

(Rs. in crores)

	1985-86	1986-87	1987-88	1988-89	1989-90	1990-91
Defence	22.7	23.9	21.4	20.8	27.8	15.7
Interest	21.6	21.6	23.4	25.7	26.4	29.9
Subsidies	13.7	13.2	12.6	14.1	16.7	12.5
Grants to States and UTs	20.3	18.4	19.4	18.1	13.2	17.5
Others	22.7	22.9	23.2	21.3	25.9	24.4
Total	100.00	100.00	100.00	100.00	100.00	100.00

(Source: *Hindustan Times*)

RICE

Rice production in the country is expected to be an all time high 72 million tonnes.

RIOT WIDOWS IN DELHI

A total of 949 flats have been allotted to widows of November riots in Delhi.

ROADS

The Working Group on Roads for the 8th Five-Year Plan has estimated that due to poor condition of roads, the losses to the economy are about Rs. 6,00 crores annually.

RUNAWAY POPULATION

When the world's population touches the six-billion mark by the end of the century, Indians will comprise on-sixth of it. The alarming population growth is the biggest challenge facing the government. That India's population growth rate, presently put at 17 million a year, may be lower than that of some other developing countries, provides little consolation in view of the current population level of 835 million.

A recent study by two population experts of the Operations Research Group at Baroda has the basis of the fertility pattern, etc., divided the country's 350 districts into 16 zones and four regions. While one region comprising 110 districts and 41 per cent of the population reflects the positive impact of family planning on the fertility rates, in another comprising 56 districts (19 per cent population) fertility rates have remained low despite hardly any FP efforts. These latter districts are in Uttar Pradesh and Bihar. The hard core area where the maximum effort is needed (117 districts with 27.2 per cent population) also comprises districts in Uttar Pradesh besides Madhya Pradesh and Rajasthan. The other region where special efforts have to be made includes parts of Maharashtra and Gujarat (67 districts with 16.3 per cent population). A region-wise approach is thus needed to correct the lacunae in the implementation of the programme. Emphasis should be on takiling the "hard core" districts which will determine the success or failure of our population control efforts.

(Source: *Hindustan Times*)

RURAL ELECTRIFICATION

See Electrification of Villages.

RURAL MARKETS

A study by the Operations Research Group (ORG) has revealed, that the rural marketing for packaged consumer products has nearly trebled from Rs.

733 crore in 1984 to Rs. 2,083 crore in 1989.

RURAL POLYTECHNIC FOR WOMEN

ONGC Mahila Samiti has begun pace-setting project with the founding of a 'rural polytechnic' in Bhagwanpur village in the Doon Valley.

The rural polytechnic opened in July, over 70 students have joined it. The objectives of this project is to promote women's education in rural areas and thereby serve as a catalyst in socio-economic development of women.

This rural polytechnic is the first such enterprise in Uttar Pradesh where rural women are being helped to gain economic stability through job-oriented courses.

In the initial phase the courses offered are stenography, textile printing, cutting and tailoring and nursery teachers' training.

The fee charged will be up to a maximum of Rs. 20 per month which will be paid to the gram sabha funds for village upliftment, especially in areas where women are involved.

Initially, the ONGC Mahila Samiti will open three rural polytechnics. The first has begun in Bhagwanpur, a village 25 km from Dehradun town. The second polytechnic would be in western Doon Valley and the third would be near Doiwala to cater to the Tehri Dam oustees who are being resettled in eastern Doon. The fourth and fifth rural polytechnics will come up in the second phase are envisaged now to be situated near Badripur in eastern Doon and in Sabarwala area on the south-western side of the Doon Valley.

A scheme for aiding the rural women in procuring their own sewing machines, typewriters and other vocational tools has also been introduced under the 'Stree Shakti Programme' of the State of India. The each one teach one concept for adult education is also being promoted in the rural polytechnic, where the regular student's will be involved in teaching their illiterate counterparts. The aim is to bring excellent education to as many villages as possible thus eliminating the necessity for rural women to come to towns for such studies.

(Source: *Hindustan Times*)

S

SAHITYA KALA PARISHAD, DELHI

The Comptroller and Auditor General of India has pulled up Sahitya Kala Parishad, a Delhi Administration body entrusted with the promotion of art and culture, for unaccounted, unused and unwarranted expenses.

Parishad purchased books and furniture for Rs. 2.77 lakh during 1986-89 for a library for which neither sufficient accommodation nor a post of librarian was made available.

Paintings worth Rs. 2.62 lakhs purchased by the Parishad from various artists up to March 1987 had been stored in a cabin without being displayed.

Parishad spent Rs. 2.48 lakh for a school contingent for Republic Day parade in 1988 and although the number of participants was only 360, the costume designer was paid for 600 costumes thereby making an overpayment of Rs. 0.55 lakh.

A van acquired at a total cost of Rs. 4.49 lakh had not been put to any use.

In 1987 the Parishad spent Rs. 5.13 lakh on provisions of facilities for a light and sound show, organised by a voluntary agencies which involved unwarranted payment of Rs. 0.89 lakh to artists.

Materials worth Rs. 1.80 lakh provided for setting up of stage were not taken back. Out of an advantage paid to the voluntary agency, amount of Rs. 29,000 remained to be accounted for.

The Parishad had not conducted physical verification of the stock and stores, in the absence of which discrepancies in stores and stock could not be ascertained.

Three language academies Hindi, Urdu and Punjabi of the Delhi Administration, did not adhere to any predetermined criteria for selection of schools of teachers and prescribed curriculum.

The Punjabi Academy incurred Rs. 102.88 lakhs on engagement of 1,207 part-time teachers during 1985-89 and the Urdu Academy Rs. 7.73 lakh on 256 teachers during 1984-89, without any predetermined criteria for their selection.

Urdu Academy teachers were teaching subjects other than Urdu as well.

Punjabi Academy incurred an expenditure of Rs. 2.40 lakh without approval of the governing council; it held receptions to delegates of international seminars like Non-aligned meet and Afro-Asian summit.

Similarly the Urdu Academy incurred Rs. 5.89 lakh on receptions or dinners at the residence of dignitaries on religious occasions not concerned with the development languages.

A substantial number of copies of publications brought out by the academies remained unsold.

Against 23,250 copies of 37 publications brought out by the Punjabi Academy, not even one copy had been sold. The Urdu Academy printed 35,361 copies of 34 publications for Rs. 9.03 lakh. Out of these 18,226 copies valued at Rs. 6.04 lakh had remained unsold. The Hindi Academy had spent Rs. 5.69 lakh on publications, had not priced them and account of stock and distribution were not kept.

SCIENCE AND TECHNOLOGY

Indo-Soviet Agreement

India and Soviet Union signed and agreement for a new programme for cooperation in Science and technology for 1990-93.

Under the programme the two countries have agreed to cooperate in 10 areas of common interest such as new sources of energy, including combined use of solar and wind energies in arid zones, polymer materials and products (foam plastics, polymer memberances), technology of production of steels and alloys: Holography, and electric equipment, reports APN. The agreement was signed here last week by India's Science and Technology Secretary V Gowariker and Deputy Chairman the USSR State Committee for Science and Technology A Zhvchenko.

Unlike the previous programmes, this one is based on the proposals of more than 20 research centres, which will be activity involved in research studies in both countries.

(Source: *UNI Release*)

SCIENTISTS' SUICIDES

In the past ten year, two scientist have committed suicide. One was employed at the headquarters of the Indian Council of Agricultural Research and the other at the Indian Agricultural Research Institute.

SCRAP IN RAILWAY YARDS

47,900 tonnes of rail scrap was lying in the various railway yards. About 1.4 lakh tonnes of scrap was disposed of for Rs. 97.33 crore in 1989-90.

SEARCHES

See Raids.

SECURITY

A sum of Rs. 7.31 crore was spent on the security of the President, Vice-President and Prime Minister during 1990-91. These VVIPs were getting special security and in 1989-90, a sum of Rs. 10.33 crore was spent.

(Source: *UNI Release*)

Security Lapses

Since the last two years, 162 security lapses were detected in the security zone of the defence headquarters. Out of these, 26 were detected in 1987, 94 in 1988 and 42 in 1989, 92 service personnel and 70 civilians were involved.

SEIZURES

The value of seizures of contraband goods including gold, silver, drugs, electronic goods effected during the last three years were Rs. 443.14 crore in 1988, Rs. 554.59 crore in 1989 and Rs. 264.62 crore in the first four and a half months of 1990.

See also Skin Seizures in Delhi; Smuggled Goods.

SEXUALLY TRANSMITTED DISEASES (STD)

The number of STD patients in the country was 13.63 lakh last year against 13.21 lakh in 1988. Five regional STD centres have been established in Delhi, Hyderabad, Nagpur, Madras and Calcutta and over 300 STD clinics were functioning all over the country to provide therapeutic services and health education to the patients.

Among the States which account for the largest number of STD patients, Maharashtra tops the list with as many as 4.26 lakh STD patients, followed by Tamil Nadu at 2.29 lakh and Gujarat at 1.27 lakh.

Among the Union Territories, the maximum number of STD patients was accounted for by Delhi at 36,972 and the minimum by Chandigarh at 397.

See also Acquired Immuno Deficiency Syndiome (AIDS).

SHARE PRICE INDEX

(Share Price Index: 1984-85 = 100)

June 17, 1991	653.5
July 1, 1991	637.1
July 5, 1991	661.7

Source: (*The Economic Times* quoted in *India Today* July 31, 1991)

SIACHEN INCIDENTS

As many as 10,365 intruders were apprehended and 464 killed in exchange of fire in the Siachen sector during the last 18 months.

SILTING

Many major reservoirs, including Bhakra in Punjab, Hirakud in Orissa and Tungabhadra in Karnataka, have lost part of their storage capacity over the years due to siltation.

Bhakra, with a life expectancy of 180 years has lost its storage capacity by 5.78 per cent over a 29-year period between 1957 to 1987, Tungabhadra with a 200 to 300 year life expectancy, has lost 14.84 per cent over a 32-year period between 1963 and 1985 and Hirakud with a 400-year life expectancy, has lost 14.5 per cent over a 27-year period between 1957 and 1984.

SKIN SEIZURES IN DELHI

The Delhi Administration had seized skins of wild animals worth Rs. 21.50 lakh during December 30, 1989 to March 2, 1990. The seizures included the skins of tigers, 29 leopards, 170 otters, 141 jackals, 123 jungle cats, 165 red foxes 6014 snakes and 30 kilogrammes of tiger bones etc.

SMALL SAVINGS

Net small savings collection, including Public Provident Fund in Post Offices, during April, 1989 to March 1990, were Rs. 7700 crores as against Rs. 5496.45 crore during the corresponding period of 1988-89, registering a growth of 40 per cent.

(Source: *PTI Release*)

SMUGGLED GOODS

Smuggled Goods Seized

(Value Rs. in Crores)

Year	Amount
1985	196
1986	218
1987	251
1988	443
1989	555
1990	223

See also Seizures.

SMUGGLING

Air India and Indian Airlines Employees

Sixteen employees of Air India and 5 employees of Indian Airlines were arrested on charges of smuggling in the three years 1987 to 1989.

SPECIAL COURTS

A total of 314 special courts had been set up in 22 States and Union Territories to try cases of atrocities on people belonging to Scheduled Castes and Tribes. Uttar Pradesh had the highest number of special courts (57), followed by Madhya Pradesh (45) and Maharashtra (30).

SPORTS

The Central Government increased the number and intensity of international cooperation activities in the field of physical Education and Sports. Cumulatively, seven Sports Protocols and 45 Cultural Exchanges were concluded. Competent coaches for special training to facilitate international standard competition were engaged for the benefit of National Teams in many disciplines of sports under these Charters of Cooperation. Overall improvement in performance and greater achievements in prestigious international sports events could be secured because of such intensive special efforts.

The Indian archers won the Asian Championship in October 1989 by dethroning South Korea, the reigning Olympic Champion. Indian weightlifters gathered an unprecedented haul of medals in Commonwealth Games, 1990. In Wrestling, India won 7 gold, 2 silver and 1 bronze medals in the Commonwealth Championships, 1989. The Indian athletes performed highly creditably in Asian Track and Field Event held in November 1989 by securing overall 2nd position after China.

An amount of Rs 2,656.21 lakh was sanctioned from January 1 to June 30 for the development of sports in the country.

During 1989-90 the actual expenditure for sports and games was Rs 6.867 lakh according to revised estimates.

While no financial assistance was given by the Government to men's cricket during the last three years, for women's cricket it was Rs 4,00,134 in 1989-90.

Asiad Song

The Asian Games Organizing Committee has chosen a song with verse by Shi Xiang and music by Shi Guangnan as the Games official song.

Noted music critic Shi Yuemeng said, "It's successful piece, a fine integration of magnificence and artistic contention."

The composer of the song Shi Guangnan, who died on May 2, 1990, was one of China's leading musicians noted for his song "On the hopeful land" among many others.

One of the Asian Games songs popular among Chinese fans is "Powerful Wind of Asia".

(Source: *Hindustan Times)*

Indian Sportsmen and Nandralone

The three Indian sportsmen E.B. Misra (javelin throw, athletics), B.P. Sirohi (kabaddi) and Jagdeep Singh (52 kg wresting) testing positive for taking a banned substance, Nandralone, face a two-year-ban unless it is proved that they did it unwittingly.

This steroid drug is similar to Stanozolol which caused the banishment from international athletics for two years of Ben Johnson.

Misra was not among those selected for the Beijing Asian Games but Sirohi and Jagdeep Singh were.

(Source: *PTI Release)*

Medals

India opened their account in the medals tally of the 11th Asian Games through the admirable efforts of two women weightlifters N.Kunjarani Devi and Chhaya Adak, who won a bronze each.

Kunjarani, a bespectacled 22-year-old, totalled 140 kg to take the third place in the 44 kg category. Later, Chhaya had a total lift of 152.5 kg to finish third in the 52-kg class.

Another Indian woman weightlifter, Jaskaran Kaur, finished fifth in the 48-kg section with a total of 127.5 kg.

All the three weightlifting titles decided today were won by Chinese girls. Indian yachtsmen Ramanjit Singh and Dalbir Singh finished first in the preliminary rounds to enter the finals of the lightweight double sculls 100 metres event.

In *kabaddi*, India registered an easy 46-15 victory over China in the round robin tournament.

(Source: C.S. Rao)

International Olympic Committee (IOC)

IOC elected a woman member to its executive board for the first time in the history of its existence since 1894.

The annual IOC session approved the nomination of former Venezuelar national equestrian champion for Isava Fonseca, 69, to the 11-member executive baord.

The session also named the seventh woman in the IOC in the person of Carol Anne Letheren, 48, a renowned judge at international gymnastics competitions, who became president of the Canadian Olympic Association earlier this year.

(Source: *AFP Release)*

Olympic Council of Asia (OCA)

The Olympic Council of Asia's general assembly today decided by a big majority to bar Iraq from taking part in the forthcoming Asian Games and to suspend that country form the OCA.

The general assembly decision to exclude Iraq came in the wake of the Iraqi invasion of Kuwait, a fellow-OCA member, on August 2 and the subsequent annexation of that country.

Not since the big rumpus over the Israeli participation in the 1962 Asian Games in Jakarta did the Asiad enveloped by a major controversy involving Iraq which has brought the Gulf on the brink of a war.

Even the International Olympic Committee's stance toward Iraq is no different from the major international political powers and that stand has now been endorsed by the OCA at its general assembly today.

The general assembly decision to exclude Iraq from the Games was announced by Mr Umrao Singh, an OCA member from India. According to Mr Umrao Singh, the assembly voted by 27 to three with five absentations and one invalid vote.

The extraordinary session was chaired by OCA vice-president Roy De Silva of Sri Lanka who later commented. "No one can condone their action. Our president has been killed, our headquarters has been burned".

OCA president and IOC member Sheikh Fahad Al-Ahmed Al-Sabah was killed during the Iraqi invasion, defending the palace of his brother, the Emir of Kuwait.

A member of the Iraqi delegation, who attended the meeting, said, "It was all a game. It was pre-arranged by the OCA. There was no discussion. We will leave tomorrow".

Sources later said the general assembly meeting began with all delegates, including the Iraqis, standing in silent tribute to the memory of Sheikh Fahad.

OCA sources said Iraqi delegate Almullah Karim requested that the assembly should "not mix sport with politics". But no other delegate spoke in favour of the Iraqi presence at the Beijing Asian Games starting on Saturday.

Iraq could muster support only from Jordan and Palestine, besides their own vote.

This is the first time a member-country has been expelled from the Games, although Israel was excluded from the Asian Games Federation, the forerunner of OCA, after the episode in the 1962 Jakarta Asiad.

Though no detail of the 'neutral' countries votes were available it is understood the two Koreas, China, Japan and one other country did not take sides in the 'dispute'.

Mr Basil Hassan, said the OCA has been "bought over by the money-power of Saudi Arabia and Kuwait and this is bad for the Olympic movement of Asia as a whole".

For India, Mr Umrao Singh and B. Sevanti Adityan attended the meeting.

Thailand's Santiparb Tejavanjija, the OCA treasurer, said "the OCA has saved the Asian Games with this historic vote".

Among the 38 members of the OCA, 36 attended the meeting. Mongolia and Afghanistan did not turn up.

Kuwaiti head of delegation Sheikh Ahmed Fahad Al-Ahmed Al-Sabah, son of the slain OCA chief, came out of the Beijing hotel conference room raising his hands in triumph.

Other pro-Kuwaiti delegates emerged flashing victory and thumbs-up signs. Chinese police struggled to control a surging crowd of reporters and cameramen.

The Thai treasurer of the OCA, Tejavanija Santiparb, said the decision meant not only that Iraq was suspended from the Asian Games starting, but that it was "suspended from the OCA until such time as the Olympic movement in Kuwait returns to normal".

Baghdad last month announced that the State of Kuwait has ceased to exist, and that the territory has become Iraq's 19th province.

An annual IOC full session endorsed an executive board communique that expressed support for Kuwait's independence in sport despite Iraq's annexation of the country.

IOC president Samaranch said winding up, "We will continue to recognise the Kuwait Olympic Committee, even in exile, and this is our official standpoint . . . We hope that as soon as possible the Kuwait Olympic Committee will be able to fulfill its function in a normal manner and in its own country."

(Source: M. Madhavan quoted in *Hindustan Times*)

SRI LANKAN REFUGEES

2,22,640 Sri Lankan Tamil Refugees arrived in India between July 1983 and August 21, 1990.

A total of 2,10,721 Sri Lankan Tamil refugees were reported to be in India on December 1989. An amount of Rs 13.10 crore was spent on these refugees.

During the last three months 3731 Sri Lankan refugees have arrived in India of which 2123 have been accommodated in Tamil Nadu and 1608 in Orissa.

The inflow of Tamil refugees from Sri Lanka was related to conditions of peace and normalcy and safety and security of Tamils in the North Eastern Province of Sri Lanka.

After the signing of Indo-Sri lanka Accord of July, 1987, 25,585 refugees were returned to Sri Lanka under the arrangements of Government of India in 50 batches between December 24, 1987 and March 31, 1989.

No decision has been taken to deport all Sri Lankan Tamil militants from India. There were 1,041 Tamil Sri Lankans in special camps of Tamil Nadu.

STEEL

SAIL has recorded a 10 per cent shortfall of targeted production for the year 1989-90. Production target for the year 1990-91 had been stepped up and fixed at 9.7 million tonnes for crude steel production and 7.7 million for saleable steel. The capacity utilisation for the year had also been stepped up to 85 per cent from 80 per cent.

SAIL had achieved a production of 7.063 million tonnes in 1989-90 as against the targeted figure of 7.81 million tonnes which was about 90 per cent. Capacity utilisation was also one 76 per cent as against the target of 80 per cent. But the production of standard tested steel showed an increase to 80 per cent from the targeted 75 per cent, he said.

Energy consumption during the year also showed a decline.

SAIL would sign a Memorandum of Understanding (MOU) with the Government for its next year's targeted production.

Government had decided to adopt a revised strategy for the second phase of modernisation of the Rourkela Steel Plant.

Government had apprised the West German Government also which had offered Rs 600 crore for the modernisation scheme and the West German Government had appreciated India's decision.

Government had decided to repackage the earlier three West German packages into smaller and more packages to encourage indigenous participation and competition in the scheme. Government had to think of this repackage scheme after opening the tenders in February 1990 which were found very high as compared to the estimate of the Public Investment Board.

The tenders showed that the three West German packages totalled Rs 3331.10 crore as against the PIB estimate of Rs 1209.06 crore. After reviewing the situation, the Government had decided to go in for global tenders once again so that more companies could bid for the tenders.

Tenders which were opened on February 20 also revealed several flaws like the West German company did not offer earnest money deposit, or agree to take on overall responsibility. A number of other details were found missing, he said.

When the West German Government was apprised of the high cost of the tenders, it also shared the view of the Indian Government that the tender prices were very high. The West German Government had also agreed to extend the same loan facility if any of its companies did for the new tenders.

Giving details of the tenders which had been received earlier, the tender for the new sinter plant package was Rs 287.34 crore with a foreign exchange component of Rs 67.76 crore as against the PIB estimate of Rs 146.82 crore. The variation between the two estimates was as much as 96 per cent, he observed.

The West German tender for the second package of new steel melting shop and continuous casting facilities was Rs 1936.61 crore as against Rs 677.48 crore approved by the PIB. For the third package for modification of hot strip mill and plate mill, the tender was Rs 1107.23 crore as against the PIB estimate of Rs 1209.86 crore, an increase of 187 per cent.

Consequent upon the decision of the Government to repackage, SAIL has been asked to issue notices within a month for prequalification of the new global tenders. The new packages were aimed at giving greater indigenous participation and would ensure that foreign exchange component was not more than 25 per cent of the total cost. Government had also made arrangements to get loans from abroad if the new bidders were not from West Germany as the West German loan was tied.

As regards Burnpur steel plant, Dastur and Company which reviewed the Japanese offer had indicated that the projected investment could be brought down by about Rs 2000 crore. The Japanese proposal was for Rs 6057 crore. SAIL had reviewed the Dastur Company proposal.

(Source: *Hindustan Times*)

There has been a loss of production of crude steel by about 3.9 lakh tonnes in 1990-91 in SAIL plants due to coking coal shortage, poor quality and blend fluctuations.

STANDARDISATION

In 1937 the Agricultural Produce (Grading and Marking) Act was passed providing for the grading and marking of agricultural (and other) produce.

Grading under Agmark is voluntary for products sold within the country, it is compulsory for many items meant for export, under the Indian Custom Act and Export (Quality Control and Inspection) Act, 1963.

Though grading and marking under Agmark is voluntary within the country, the amendment made to the AP (G and M) Act in 1986 has given suitable powers that if need be compulsory grading for certain articles may be made in public interest as well as for the protection of the consumers.

Grading under Agmark is done for a number of commodities like foodgrain and allied products-paddy, rice, wheat, jowar, peas, black gram, barley, maida, kabuli gram, lentil.

Consumers can write to the Deputy Agricultural Marketing Adviser in charge of the Northern Region at 4/20 Asaf Ali Road in case they have any complaint regarding Agmark products.

In the year 1987-88 the value of commodities graded under Agmark for export was to the tune of Rs 661.65 crore and during 1988-89 it is estimated to be around Rs 725 crore.

The total values of commodities graded under Agmark for internal consumption amounted to Rs 883.16 crore in 1987-88. It is estimated to be around Rs 920 crore during 1988-89.

(Source: Satyam Mohapatia quoted in *Hindustan Times*)

SUBMARINES

The Comptroller and Auditor General has indicted the Defence Ministry for the extraordinary delay leading to heavy cost overruns in the indigenous production of the controversial HDW SSK class killer submarines.

The times required for the construction of the submarines at Mazgaon Docks Limited (MDL) had been grossly under-estimated and it had jumped up from 42 months to 81 months. This had consequently led to tripling of costs from the estimated Rs 196.91 crore to Rs 334.06 crore.

The Government has paid Rs 172.48 crore for the two sail-away submarines contracted in November 1981. Also, there was a significant industrialisation of the shipyard effort involved in indigenous construction. Originally estimated at Rs 63 crore, this effort is now estimated to cost Rs 152 crore.

Due to the delay on the part of the Ministry in taking a decision about the future construction plan, the facilities created at MDL at a cost of Rs 44.7 crore are under-utilised. Machinery costing Rs 12.69 crore will remain idle after the completion of the two submarines. There was a mismatch between the induction of submarines and the commissioning of related training simulators, and non-availability of certain essential training equipment.

A number of items of work on the submarine could not be attended to either for want of spares or owing to delays in creating the facilities required.

An incorrect assessment of spares at the contract finalisation stage resulted in the procurement of Base and Depot spares at a higher cost of Rs 91 crore against an ad hoc provision of Rs 32 crore.

There was a "mismatch: between the induction of the submarines, receipt of their torpedoes and the creation of facilities for their storage and two torpedoes, each costing lakhs of rupees, were found unserviceable on receipt. When the naval armament depot (NAD) received the torpedoes the major portion of their warranty period had expired. Seventeen more of these torpedoes were found defective and the Ministry had intimated that they were under repair free of cost by the manufacturers.

The project envisaged the construction of workshops, buildings and dry dock in MDL, including an augmentation of the existing facilities.

The estimated cost of these facilities was Rs 41.99 crore to be completed by January 1984. The facilities were actually completed by September 1984 at a cost of Rs 41.80 crore. Since an order for the construction of further submarines had not been placed on MDL, the facilities created at MDL would remain idle after the completion of the construction of the two submarines.

Two mine saddles for carrying mines were procured along with the two submarines at a cost of Rs 5.87 crore from firm "A" for use by four submarines. Naval Headquarters evaluated several mines and shortlisted two foreign firms "G" and "H".

SUGAR INDUSTRY

Prices

Sugar prices remained fairly stable during the past few months.

Levy Quota

The Government decided to allow five per cent ad hoc increase in the monthly levy sugar quotas to States and Union Territories.

Production

Sugar production reached a new high of 105.59 lakh tonnes in the October-May period of the sugar year 1989-90. The production in the corresponding period last year was 85.78 lakh tonnes.

During the month of May alone, 7.82 lakh tonnes of sugar were produced, as against 1.42 lakh tonnes in May 1989.

With a total offtake of 67.47 lakh tonnes, including 1.82 lakh tonnes of imported sugar, in October-May period, the closing stock stood at 52.77 lakh tonnes (including) 0.06 lakh tonnes of imported sugar) on 31st May 1990.

(Source: *PTI Release)*

India created a world record in sugar production this year. It is the first country to have crossed the 100 lakh tonnes production mark.

The production was 101.64 lakh tonnes while the total production by the end of the current season is expected to be around 106 lakh tonnes-nearly — 20 lakh tonnes more than last year's production.

Not only has India become the first country to achieve 100 lakh tonnes production but it has surpassed its own record of 91.10 lakh tonnes which it had achieved in 1987-88. The highest production of sugar from cane in the world, other than India, was achieved by Brazil in 1983, when it produced 88.84 lakh tonnes.

The increased level of sugar production in the current season is due to the incentives given by the Government to the sugar factories for increased crushing. The Union Government has also announced incentives for early crushing next season and expects similar higher production in the coming years.

(Source: *Hindustan Times*)

SUPREME COURT AND HIGH COURTS

Cases

As many as 39,485 cases are pending in the Supreme Court — 6,480 of them for over 10 years.

Besides increasing the judge strength, various steps like grouping of cases involving common question of law, constitution of special benches have been taken by the Government for speedy disposal of cases.

Government has also constituted a committee of three Chief Justices of High Courts to make an indepth study of problem of arrears in courts and to suggest remedial measures.

3,90,480 cases were pending in the Allahabad High Court — 30,245 of them for over 10 years.

As many as 1,08,379 cases were pending before Supreme Court as on June 30, 1990. Of this 44,815 were regular hearing matters and 63,564 admission matters.

Cases disposed by the Supreme Court were 48,118 in 1989 as against 44,252 in 1988 and 46,132 in 1987. High Courts disposed of 7,70,946 in 1989, 7,14,928 in 1988 and 7,00,407 in 1987.

Vacancies

Three vacancies of judges had been filled in the Supreme Court since December 1 this year. There are two more vacancies of the judge in the Supreme Court.

T

TV

See Doordarshan.

TAXATION

Arrears

As on end-December 1989, the amount of money which Government had to recover from industrialists and businessmen was Rs 305.31 crore in Central excise, Rs 943.23 crore in Customs and Rs 171.34 crore in Income-tax including corporation tax.

Collection

Provisional collection figures for direct and indirect taxes up to Nov. 30 last was Rs 3,442. 97 Core and Rs 27,327 crore respectively

Share

The share of direct taxes had gone down from 22.80 per cent in 1989-81 to 19.25 per cent by 1989-90, while that of indirect tax had gone up from 75.36 per cent between these 10 years.

TEA

Export

Export of tea during 1990-91 are estimated to have touched Rs. 1045.41 crore as against Rs. 862.45 crore during 1989-90. The unit value realisation is estimated to have gone up to Rs. 52.59 per kg during 1990-91 from Rs. 43.18 per kg in 1989-90. In terms of quantity tea exports amounted to 199.48 million kgs as against 202.81 million kg in the previous financial year 1989-90.

(Source: *Yojana*, July 15, 1991)

TECHNOLOGY INFORMATION FORECASTING AND ASSESSMENT COUNCIL (TIFAC)

Department of Science and Technology has set up a Technology Information Forecasting and Assessment Council (TIFAC) to facilitate forecasting and assessment of the indigenous and foreign technology and build up appropriate information centre.

TECHNOLOGY MISSION ON DRINKING WATER

The Comptroller and Auditor General of India had criticised the functioning of the technology mission on drinking water which has failed to achieve its objective of supplying potable water to 1,37,155 villages by the end of the Seventh Plan.

The mission was to undertake conservation measures for sustained water supply. For this, 55 mini-mission (pilot project) were taken up. Five Sub-Missions under different nodal agencies were also set up to deal with the problems of source finding and quality of water, control of fluorosis, removal of excess iron, eradication of guineaworm and control of brackishness.

The mission was to assure supply of required quantity of potable water within a reasonable distance to say 1.6 km and within a depth of 15 metres in 1,37,155 problem villages. Though 1,28,823 problem village were covered during the period 1986-90 against 1,37,155 problem village as envisaged to be covered, reported coverage included partially covered villages also. At the end of March 1990, there were 1,18,163 partially covered villages including those pertaining of the pre-mission period. The mission had no information about the quality of potable water per capita per day being supplied and percentage achievement in case of installation of hand pumps as against the norms of 40 litres per capita per day or one hand pump for every 250 persons.

Quality monitoring of water supply did not receive the intended emphasis. For testing quality of water, only 33 laboratories could be set up against the target of 100 to be operational by November 1989.

Against the target of 1,825 villages, to be tackled for excess flouride in nine States, only 796 villages were covered.

Likewise, six defluoridation plants were commissioned even though 130 plants were targeted.

For eradication of guineaworm in affected villages, 6,754 villages were covered till December 1989 against 9,238 villages identified for the purpose in six States for alternative source of water or conversion of step wells into sanitary wells. Out of 5,578 step wells to be converted into sanitary wells, only 2,576 wells were converted.

For rural Water Supply Schemes during the seventh Plan period expenditure totalling Rs 4,385.30 crore was incurred by the States and Union Territories.

The mission approved schemes for Rs 210.18 crore and released about Rs 120 crore against which the expenditure incurred was Rs 111 crore during 1986-90. In addition, Rs 26.86 crore were released for activities of the sub-mission; details of expenditure were not available with the mission.

There was significant shortfall in the implementation of schemes for water conservation. Against the outlay of Rs 75 crore, Rs 17.66 crore were released and the actual expenditure was Rs 6.15 crore only. Utilisation was less than 50 percent of the release in several districts of seven States. No funds were released for water conservation works in nine States.

Funds to the extent of Rs 300 lakh were diverted for works outside the purview of the mission.

Monitoring of the programmes at district level and State level was deficient. The monthly progress reports received from the State project coordinators depicted financial and physical targets only. As such the Mission could not carry out the intended detailed monitoring.

The agencies for evaluating the progress of the individual project annually and to suggest methodology and approach for rectifying deficiencies wee not identified as envisaged. As a result, evaluation of the implementation of the mini-mission projects and activities of sub-missions had not been done.

(Source: *Hindustan Times*)

TELECOM DEPARTMENT

Borrowing

The total market borrowing by the Telecom Department as on March 31, 1989 was Rs 1550 crore excluding MTNL. The liability during 1990-91 is expected to be Rs 174 crore, he said.

TELECOM POLICY (DRAFT)

The draft National Telecom Policy that outlines the measures to be taken no take the country forward in this vital sector, identifies 10 thrust areas as reasonable cost of telecom services, increased access to the public telecom network in both the urban and rural areas, appropriate technology for the hilly, tribal, and remote areas, special facilities for handicapped persons, and higher production of telecom equipment.

The other areas are liberalisation of licensing policy to facilitate production in the private and joint sectors, modernisation of telegraph network, R & D development with increased investment, introduction of modern technology not only for switching and transmission but also for data communication, mobile communication, video transmission, voice mail, etc, and better measures for satisfying customers.

The policy aims at achieving "self reliance" in telecom technology. It seeks to develop and integrated digital network from the present analogue network. Later an integrated services digital network would be developed in a phased manner. Villages too would be covered after a telephone had ben provided in each panchayat.

Government proposed to run telecom services on a commercial basis so that a major part of the expansion and upgradation of telecom services could be financed from internal resources.

Telecom tariffs would be determined on commercial basis, but it would differentiate between different classes of customers and appropriate cross-subsidisation would also be done. There would be separate rates for commercial and non-commercial users.

Telecom infrastructure was capital-intensive. It has potential for high return. The Government wanted to step up indigenous production and development to conserve foreign exchange.

The policy provided for consolidating the present production base, standardising telecom products, modernising and expanding the capacity in both the public and private sector, ensuring quality of products and services, and facilitating exports.

Quality would be improved at the component stage so that the finished product became automatically of high quality. There would be progressive liberalisation of licensing policy.

The policy also aimed at improving human resources management to improve productivity. A major part of the expansion and modernisation would be financed from internal resources, but market borrowings would also be resorted to in case of requirement.

As regards postal services, norms were being evolved for expanding postal services in the rural areas. Automation and mechanisation were being introduced in postal sorting in the four metropolises.

(Source: *Hindustan Times)*

TELEVISION

See also Doordarshan

TELEPHONES

Connections

Over 46 lakh telephones function in the country as per official data. The draft 8th Plan envisages net addition of 52 lakh new connections.

(Source: *Yojana)*

As many as 4,58,553 telephone connections were provided in Delhi as on January 31, 1990. It was proposed to clear the waiting list of South Delhi during the 8th Five-Year Plan progressively subject to availability of equipment

Electronic Exchanges

The target fixed for opening electronic exchanges during 1990-91 was 2,957. The number of electronic exchange opened during the year was 2987. During 1991-92, 4,472 exchanges are proposed to be opened. There were 3727 electronic exchanges up to June 1990.

Faults

The daily average number of faults in telephones in Delhi is 4,000 and it increases to 6,000 during the rainy season. There are 4,66731 telephones in Delhi as on July 1, 1990.

C-Dot Switches

The C-Dot has submitted a bid and signed a letter of intent with a US firm M/s Motorola to enter into a contractual relationship that would enable Motorola Cellular system to work with C-Dot switches through appropriate interfaces.

TERRORISM AND TERRORISTS

Killings

A total of 1,210 persons were killed in Punjab due to terrorist violence up to july this year.

See also Punjab killings.

Kidnappings

Terrorists in Punjab have adopted yet another technique of raising funds and extorting huge amounts of money from innocent citizens living in the border district of the State.

They take away some female members of a family at gun-point and keep them in their custody for two or three days.

(Source: *Hindustan Times*)

Damage to Railways Property

The damage to railway property in 1989-90 in Punjab due to terrorist activities was about Rs 5.28 lakh. During 1988-90 it was about Rs 16.80 lakh.

Delhi Scene

Terrorist attacks, robberies and dacoities are on the increase in the Capital. In the first six months of 1991, 18 dacoities, 135 robberies and 14 terrorist attacks wee registered while in the corresponding period (Jan. 1, 1990 to June 30, 1990) of last year 12 dacoities, 116 robberies and 12 terrorist attacks were registered.

277 people, including 14 terrorists, were arrested in the cases from Jan. 1,

1990 to June 30, 1991. 10 people killed and 87 injured in terrorist attacks in the Capital from April to June. The Delhi Administration had paid Rs 3,24,000 as compensation to the next of kin of those killed and to those injured.

Increased patrolling, posting of pickets at strategic points, strengthening of the intelligence apparatus, coordination meetings with officials of neighbouring States and formation of anti-terrorist cells in each police district are among the measures taken by the Government to arrest this trend.

TEXTILE EXPORT POLICY

See under Export.

TIGERS

There has been a noticeable decline in the population of tigers in the north-eastern region, which was otherwise regarded as a natural reserve for tigers because of the dense forest suitable as a natural habitat for the tiger.

Central Government has been increasingly providing funds for the maintenance of tiger reserves in the country, and last year's provision was of Rs 6.5 crore. Besides, the State Governments spent about Rs 3.2 crore out of their own resources on tiger reserves in their areas.

The Central Government also extended 100 per cent assistance to the States for non-recurring expenditure on tiger projects.

The male-female ratio of the tiger population was 40:60 (female tigers far outnumbering the males). The latest census figures of the tiger population showed that in Meghalaya, the population of the big has dropped from 125 to 34, in Mizoram, it fell from 33 to 18, and in Arunachal Pradesh, the decline was more steep, from 217 to 135. Tripura forest earlier had five tigers, but even that disappeared somehow and the population now recorded was "nil".

About 75 per cent of the tiger population in the country preferred to stay in the tiger reserves, and the rest in the wild forest — unprotected areas. There has however been an overall increase in the population of tigers, from 4,005 in 1984 to 4,334 in 1989.

While there are no official figures available about the incidence of poaching in reserve forest areas, five forest officials were killed (by poachers?) while chasing illegal hunters in the forest areas during the last year. The Government paid *ex gratia* to the bereaved families. The Government has also introduced a Bill in Parliament to amend the Wildlife (Protection) Act in order to provide a more stringent punishment to illegal poachers — the imprisonment would be three years, as against two years now.

A steering committee under the chairmanship of the Prime Minister has also been set up to review and monitor the maintenance of tiger reserves in the country. The total area covered by the 18 tiger reserves was 28.017 sq km — the largest area being in Nagarjunasagar reserve in Andhra Pradesh (3568 sq km). The Kanha tiger reserve in Madhya Pradesh was spread over 1945 sq km, Manas in Assam covered 2840 sq km, and Corbett National Park was spread over 521 sq. km.

The largest number of tiger population was in Madhya Pradesh (1985), followed by 735 in UP, 417 in Maharashtra, 376 In Assam, 352 in West Bengal and 243 in Orissa.

Central Government also provided an additional grant of Rs 11 lakh to Orissa for the relocation of people living inside the core area of the park to places outside the park. The same problem was faced by the managers of the Kanha national park. (The number of tigers in Corbett also recorded a drop from 92 to 70 at the time of the 1989 census).

A nature interpretation centre with US assistance was set up in Kanha, and the project has since been completed. The object of the project was to generate interest in wildlife among the youth, particularly the students to create greater awareness for the eco-system.

(Source: *Hindustan Times*)

Sunderbans Tiger Reserve

The number of tigers in the Sundarbans tigers reserve had increased from 264 in 1984 to 269 in 1989. There were no indications that the tigers faced a food or space crisis.

TOBACCO

Tobacco exports declined from an all-time high of 132 million kg in 1981-82 to 73 million kg in 1989-90. The demand for tobacco was going down owing to the anti-smoking campaign the world over.

TRADE DEFICIT

Notwithstanding the excellent export performance during 1989-90, recording a 36.5 per cent growth over the previous year, the overall trade deficit stood at Rs 7,730 crore, thus showing a marginal rise of 4.3 per cent.

India's exports in 1989-90 have now been provisionally estimated at Rs 27,681 crore as against Rs 20,281 crore the previous year, while imports also went up to Rs 35,412 crore from Rs 27,693 crore in the same period.

Because of the huge import bill on account of the purchase of the A 320 Airbus and also the massive petroleum imports, the trade deficit, rose by 4.3 per cent otherwise it would have actually recorded a decline of 60.5 per cent. The purchase of 13 A-320 Airbus amounted to Rs 758.96 crore, while petroleum imports totaled Rs 6,192 crore. The petroleum import figure, it is pointed out, represents a 48.2 per cent increase compared to the Rs 4,178 crore on this account in 1988-89.

Excluding the Airbus and petroleum import figures, the country imports would have represented an increase of only 21 per cent in rupee terms as opposed to the 27.9 per cent when these two vital imports are taken into account.

On the export front, the figure of Rs 27,681 crore during 1989-90 represents a 36.5 per cent growth in rupee terms over the figure of Rs 20,281 crore in 1988-89. The growth in terms of the US $ is put at 18.7 per cent.

This export growth rate is said to be the highest in the last 22 years. The previous best export performance was in 1966-67, when a 42.8 per cent export growth was witnessed. The export target for 1990-91 is set at Rs 36,000 crore.

The acceleration in export growth in the last three yeas has not only been sustained but bettered considerably each successive year. The export growth has thus gone up from 14.3 per cent in 1986-78 to 25.4 per cent in 1987-88, to 30 per cent in 1988-89 and to 36.5 per cent in 1989-90.

The export figure of Rs 27,681 crore in 1989-90 is two and a half-times that of the performance during 1985-86, when the export turnover was of the order of Rs 10,895 crore. Incremental export in the whole of Sixth Plan (1980-85) were put at US $ 1,650 million, whereas in 1989-90 they have been of the order of $ 2,622 million over that of the previous year. In dollar terms, the export have nearly doubled in these five years.

The export-import ratio, that is the extent to which exports finance imports, is said to have steadily risen from 55 per cent in 1985-86 to 70 per cent in 1987-88 to 72 per cent in 1988-89 and to 78 pr cent in 1989-90.

Export during 1989-90 have not quite achieved the target of Rs 28,025 crore. The main reason for this is said to be the sizeable shortfall in gem and jewellery exports that slipped by about Rs 1,000 crore.

(Source: *Hindustan Times)*

TRAFFIC

A three-day workshop organised in Delhi recently by the Delhi Police and the Kerala State Electronics Development Corporation to discuss modern traffic management schemes requiring minimum financial commitments has highlighted the magnitude of the problem and total chaos prevailing of roads in different metropolitan cities. Broad recommendations made at the workshop assume greater significance particularly because of the rapid urbanisation, population growth, the increase in mobility and liberal policy in regard to the manufacturing of automobiles and their rapid growth on roads, causing greater road safety hazards for lakhs of people.

The faster pace of urbanisation in metropolitan cities and other developing towns has given a boost to the automobile industry. While the number of vehicles in the by towns has increased manifold, the road expansion programmes have remained either neglected or given least priority.

Transport problems have reached a crisis point. The increase in the number of accidents, the operational cost of vehicles, increase in pollution and overall degradation of environment have been a matter of serious concern for the experts and concerned official agencies.

In Delhi from less than two lakh vehicles in 1970, the number of automobiles in the Union Territory had gone up to more than 16 lakh last year. The major concern for the road planners and traffic experts, particularly in the metropolitan cities, has been the absence of the corresponding increase in road expansion programmes.

In the process of rapid urbanisation, increase in population and need for fast movement, the traffic scenario of Delhi had become alarming. Two lakh new vehicles were added to the Capital every year.

In another 10 years, the number of vehicles in the Capital was likely to go upto 35 lakh from the existing 17 lakh.

The rate of accidents in Delhi has shown a steady decline since 1978. While total accidents increased by 12.3 per cent from 1982 to 1983 and thereafter continued to increase during the subsequent years, except in 1986 and 1987, the rate of accident per 1000 vehicles during these years especially since 1984 and thereafter showed a steady decline by 0.43 per cent, 0.37 per cent, 0.96 per cent, 0.82 per cent, 0.24 per cent and 0.22 per cent respectively. During 1988 the fatal accidents shot up by 15.27 per cent against the number of fatal accidents in the previous year which rising trend dropped to 7.26 per cent during 1989.

The number of motor vehicles registered in Delhi had been showing a rapid growth. In 1989 the total number of vehicles registered stood at 16,31,617 as against 14,15,931 in 1988. The total number of vehicles 10 years ago, i.e., 1979 was only 4,79,272. The vehicular population in Delhi has more than trebled that of 10 years ago. This clearly shows the inadequacy of the Mass Transportation System in coping with the required needs.

Nearly 30 per cent of the estimated 16 lakh vehicles in the Capital failed to conform to the specified pollution standards.

(Source: A.R. Wig in *Hindustan Times)*

See also Accidents: Road.

U

UNEMPLOYMENT

As many as 332.27 lakh job-seekers, all of whom are not necessarily unemployed, were on the live register of employment exchanges in the country at the end of February 1990.

There were as many as 30,869 medical graduates and post-graduates on the live register of the employment exchanges in the country as on December 1989.

The estimated numbers of educated and semieducated (literate and primary and middle school educated) and illiterate usually unemployed persons aged 15 years and above in the country as on January 1, 1988 were about 3.7 million and 3.3 million and 1.8 million respectively.

Punjab Situation

The Punjab Employment Department has registered 2,03,857 persons seeking employment and against this 8,145 persons placed in the various government/semi-government departments and public undertakings in the state during 1988.

6,124 persons were motivated to appear for the competitive examinations and 1,125 came up for self-employment schemes. A sum of Rs. 65.60 lakh has been disbursed as unemployment allowance to 9,772 candidates. There were 5,84,643 persons on the live registers as on December 31, 1989.

(Source: *Hindustan Times*)

UNITED LIBERATION FRONT OF ASSAM (ULFA)

ULFA Killings

213 people had been killed by ULFA in 1988, 1989, 1990 and up July 18 during the current year. The year-wise break-up is: 1988 — 24; 1989 — 28; 1990 — 89; 1991 (upto July 18) 72.

V

VIP FOREIGN TRAVELS

Foreign tours of ministers as well as senior officials would be substantially curtailed in the current year. The expenditure on foreign travel during 1990-91 should be limited to 75 per cent of the actual expenditure.

VIJAYANTA TANKS

The Army has come under sharp attack for the inordinate delay of six years in the modernisation project of upgrading the Vijayanta and Soviet T-55 tanks which led to heavy cost over-runs and a setback to defence preparedness.

The project which was to be completed by 1990-91 would now only be completed by 1996.

The Army had mooted the modernisation project because of the delay in the development of indigenous Main Battle Tanks (MBT) project, christened "Arjun".

The phasing out of the Vijayanta and T-55, which was to start in 1986, was cancelled and plans mooted to upgrade these tanks to convert them into contemporary tanks for battlefield of nineties and beyond.

The basic idea of modernising the tanks was drawn up in 1979-80, but comprehensive sanction for the project was not issued even till October 1990. It was ruled belatedly in August later in 1990 that the expenditure on modernisation would be met from the maintenance Budget.

The sanctions for different components were piecemeal and nodal workshops for undertaking the modernisation programme were not equipped till October 1990. As a consequence sanctions issued in July 1986, at Rs. 1.97 crore could not be operated.

Due to reduction in the scope of modernisation at a later stage in July, 1988, there were excess provisioning of passive night vision devices and financial repercussions of Rs. 18.55 lakh on account of short closure of indent.

Despite failure in trials held in 1982, the report pointed out, simplified fire control systems were imported at Rs. 8.7 crore under the modernisation programme and these were later "found to be defect prone and non-functional".

For want of sanction for procurement of engineering package required for maintenance cover, "tank fire control systems IA and IB costing Rs. 28.80 crore could not be exploited fully.

Though final decision in selection of new power pack for repowering the tank was still awaited till October 1990, expenditure to the tune of Rs. 66.21 crore had been incurred for supplies received up to April 1990. Other sophisticated components worth over Rs. 47.19 crore received under modernisation programme were not fitted in the tanks and remained unutilised till October, 1990.

Under the modernisation programme, the report said, a specified number of about over 300 to 500 Vijayanta and T-55 tanks were to be upgraded at a cost of Rs. 518.20 crore. The modifications to be carried out included providing new gun control system, low level light television and "Kanchan armour".

The "Kanchan armour", which is futuristic in conception, has been developed for use of main battle tank. It was necessary to upgrade these tanks as the Army wanted to retain a reliable tank fleet after 1995. During January to June, 1990 money suply recorded an increase of 7.3 per cent as against 6.9 per cent increase in the corresponding period of 1989.

(Source: *Hindustan Times)*

VOTERS

Electorate in States and Union Territories

State/U.T.	Men	Women	Total
A.P.	21,433,197	21,314,665	42,747,682
Arunachal Pradesh	279,308	239,921	519,229
Assam	6,323,078	5,538,279	11,861,357
Bihar	28,124,164	24,521,071	52,645,235
Goa	389,054	379,019	768,073
Gujarat	12,801,439	12,070,532	24,871,971
Haryana	5,233,092	4,464,311	9,679,403
H.P.	1,554,891	1,518,035	3,072,926
Karnataka	14,724,709	14,010,406	28,735,115
Kerala	9,688,768	9,920,177	19,608,945
M.P.	19,432,452	18,266,911	37,699,363
Maharashtra	25,434,433	23,230,322	48,664,755
Manipur	615,188	616,455	1,231,643
Meghalaya	474,945	466,617	941,562
Mizoram	208,480	205,347	413,827
Nagaland	455,476	359,260	814,836
Orissa	10,606,570	9,191,028	19,797,598
Punjab	7,077,864	6,029,160	13,107,024
Rajasthan	14,041,298	12,445,948	26,487,246
Sikkim	104,799	96,440	201,239
Tamil Nadu	20,165,453	19,660,956	39,826,409
Tripura	796,288	761,930	1,558,218
U.P.	44,190,361	36,012,524	80,202,885
W.B.	22,107,328	19,242,852	41,350,180
A. & N.I.	93,642	68,089	161,731
Chandigarh	202,365	154,702	357,067
Dadra and N.H.	38,086	36,870	74,956
Daman and Diu	28,713	28,57	57,285
Delhi	3,309,145	2,719,539	6,028,684
Lakshadweep	16,135	14,578	30,713
Pondicherry	305,350	285,693	591,043
Total	270,256,171	243,870,209	514,126,380

(Source: *UNI Release*)

W

WAFK PROPERTY IN DELHI

Orders for implementation of the Burney Committee report on Wakf properties in Delhi were issued in March 1984 but these could not be implemented due to a stay order passed by the Delhi High Court.

WOMEN

Health

A survey conducted by the Indian Council of Medical Research under the National Nutrition Monitoring Bureau (NNMB) during 1972-79 in 10 States revealed that about 24 per cent of the women had low body weight and height.

As a result of such undernourishment these mothers fall into high risk categories and are likely to suffer from obstetic complications.

(Source: *PTI Release*)

Crimes against Women

As many as 117 cases of rape, 1,362 cases of eve-teasing, 105 cases of molestation and 69 cases of dowry deaths have been reported in Delhi during the last eight months.

While incidents of rape and kidnapping of women and girls in the country had declined in 1989 in comparison to the previous two years, the number of molestations and dowry deaths had increased.

7,767 rape cases were reported in 1987, 8,706 in 1988 and 8,207 in 1989. The number of kidnappings was 9,016 in 1987, 9,633 in 1988 and 9,202 in 1989.

The number of molestations totalled 16,292 in 1987, 17,836 in 1988 and 18,437 in 1989, while dowry deaths numbered 1,912 in 1987, 2,209 in 1988 and 3,829 in 1989.

In Delhi, the number of molestation cases in 1989 was 159, rape (152), dowry deaths (110) and cruelty by husband or in-law (336). The figures for 1988 were molestation (130), rape (127), dowry deaths (103) and cruelty by husband or in-laws (349).

Among crimes against women during 1990 were 9,517 cases of rape, 20,186 of molestation and 11,689 of kidnapping and abduction.

Altogether more than 43,700 cases of crime against women, including rape molestation and dowry and death were registered in 1990.

Dowry Deaths

More than 11,000 women were killed in three years between 1988 and 1990 in dowry-related murders and suicides in the country. Nearly 28,000 cases of rape were registered in the last three years.

As many as 878 cases of dowry murders and 1,479 cases of dowry suicides were registered in the country in 1990. A total of 2,478 cases of death under Section 304-B of the IPC were also registered during the 1990.

Women Criminals in Delhi

A total of 1,890 women criminals were arrested in the Capital until November 30, 1990 on charges of murder, attempt to murder, rioting, burglary, theft and other offences.

The number of women arrested in 1987 was 2,000, in 1988 was 1,605 and in 1989 was 2,327.

WORLD RECORD

Twenty-six year old Ananth Kumar by etching 515 letters on a single grain of rice has his name entered in the Guinness Book of World Records.

An engineer with the Karnataka Power Corporation Ananth Kumar, who hails from Mysore, told UNI that he had achieved the feat in a single day in the presence of several dignitaries. The letters were written without the use of magnifying glass.

The previous record was 437 letters on a grain of rice according to the certificate and citation received by Mr Ananth Kumar.

(Source: *UNI Release*)

Y

YAMUNA BRIDGE

Yamuna bridge, near ISBT, connecting trans-Yamuna areas with rest of Delhi, besides providing easy and direct link to UP, Punjab, Haryana, Jammu and Kashmir, Himachal Pradesh and northern region of the country was opened for light traffic.

The bridge, costing Rs. 38 crore, is being constructed by the PWD Delhi Administration. The 3.7 km long bridge has eight-lanes and when completed will carry 1.25 lakh passenger car unit per day. The construction started in 1984 and was originally scheduled to be completed in three year.

(Source: *Hindustan Times*)

YOUTH BODIES

A directory of youth organisations published in 1981 has listed total number of approximately 547 youth organisations in the country. During 1989-90 the voluntary youth organisations had received assistance of Rs. 2,13,87,131 while in 1987-88 and 1988-89 it was Rs. 1,57,48,436 and Rs. 2,42,91,314 respectively.

Z

ZINK

Aguch-chanderiya Integrated Lead-Zink Project.
See Mines and Mining: Implementation of Integrated Lead-Zink Project of HZL.

CICIL LIST

CONCEPTS IN COMMUNICATION, INFORMATICS & LIBRARIANSHIP

(CICIL)

General Editor: **S.P. AGRAWAL**

CICIL-1 Development of Library Services in India: Social Science Information
S.P. Agrawal Rs. 150

CICIL-2 Development of Documentation in India: Social Science Information
S.P. Agrawal Rs. 250

CICIL-3 UNESCO and Social Sciences: Retrospect and Prospect
S.P. Agrawal and **J.C. Aggarwal** Rs. 250

CICIL-4 INDIANA : A Bibliography of Bibliographical Sources
M.K. Jain Rs. 200

CICIL-5 Dimensions of Library and Information Science: Kaula Festschrift
V. Venkatappaiah (Eds.) Rs. 600

CICIL-6 Book Numbers: Some Indian Methods
M.P. Satija and **S.P. Agrawal** Rs. 100

CICIL-7 National Policy for University Libraries in India: Problems and Perspectives
N.B. Inamdar and **L.S. Ramaiah** (Ed.) Rs. 250

CICIL-8 Media Utilisation for the Development of Women and Children
B.S. Thakur and **Binod C. Agrawal** Rs. 150

CICIL-9 Terrorism: An Annotated Bibliography
Susheela Bhan (Ed.) Rs. 275

CICIL-10 Transnational Library Relations: The Indo-American Experience
M.B. Konnur Rs. 175

CICIL-11 Geography of Marketing and Commercial Activities in India (Documentation on Research Information)
R.S. Dixit Rs. 130

CICIL-12 Information India 1988: Global View
S.P. Agrawal and **J.C. Aggarwal** Rs. 200

CICIL-13 Women's Education in India: Historical Review, Present Status and Perspective Plan
S.P. Agrawal and **J.C. Aggarwal** Rs. 360

CICIL-14 Cultural Communication in India: Role and Impact of Phonograms
J.M. Ojha Rs. 220

CICIL-15 Second Historical Survey of Educational Development in India: Select Documents 1985-1989
S.P. Agrawal and **J.C. Aggarwal** Rs. 490

CICIL-16 Documentation Encyclopaedia of UNESCO and Education
J.C. Aggarwal and **S.P. Agrawal** Rs. 800 (set of 2 parts)

CICIL-17 Attitude Towards Information: A Study of Post-Graduate Teachers and Researchers in Social Sciences
N.G. Satish (in press)

CICIL-18 Librarianship and Bureaucratic Organisation: A Study in the Sociology of Library Profession in India
P.K. Jayaswal Rs. 175

CICIL-19 Literature Search
K.A. Isaac Rs. 150

CICIL-20 Child Education in India: Index to Scholarly Writings in Indian Educational Journals and Newspapers since independence
S.P. Agrawal and **Naresh Kanta** Rs. 100

CICIL-21 Lok Sabha and Vidhan Sabha Elections in India 1989 & 1990: Process and Results with Comparative Study of Manifestos
S.P. Agrawal and **J.C. Aggarwal** Rs. 250

CICIL-22 Islamic Studies in India: A Survey of Human, Institutional and Documentary Sources
Mohamed Taher Rs. 180

CICIL-23 Panchayati Raj: An Annotated Bibliography
N.B. Inamdar, N.G. Satish and **Anil Takalkar** 290

CICIL-24 Information India 1989-90: Global View
S.P. Agrawal and **J.C. Aggarwal** Rs. 500

CICIL-25 First Handbook of Psychological and Social Instruments
Udai Pareek and **T. Venkateswara Rao** Rs. 500

CICIL-26 Educational and Social Uplift of Backward Classes: At What Cost and How? Mandal Commission and After
S.P. Agrawal and **J.C. Aggarwal** Rs. 275

CICIL-27 Mass Media and New Horizons: Impact of TV & Video an Urban Milieu
H.R. Trivedi Rs. 180

CICIL-28 Indian Writings on Education : An Indicator to Indian Educational Journals Grouped by Subject Deseriptors 1979 to 1986 (in two parts)
S.P. Agarwal Rs. 900 (per set)

CICIL-29 Nation in Crisis
S.P. Agrawal and **J.C. Aggarwal** Rs. 200

CICIL-30 Information India 1990-91: Global View
S.P. Agrawal and **J.C. Aggarwal** Rs. 500

CICIL-31 Development of University Libraries in India after Independence
O.P. Gupta Rs. 270

CICIL-32 Handbook on Child : With Historical Background
Pramila Pandit Barooah Rs. 320

CICIL-33 Weaker Section in Indian Society : Undiscovered Past, Uncertain Present and Unpredictable Future : Index to Writings in Indian Scholarly Journals and 10 National Newspapers Classified under 1500 Descriptors 1886-1990
S.P. Agrawal and **Rajeev Kumar Sharma** Rs. 600

CICIL-34 Communication Informatics and Librarianship in India : An Introductory Guide to Publications Based on a Survey of Libraries in Capital City of India
S.P. Agrawal and **Pushpa Rani Sharma** Rs. 720

CICIL-35 Problems of Collection Development in Special Libraries
M.S. Sridhar Rs. 225

CICIL-36 National Information Resources for Social Sciences in India : Mobilisation, Dispersal and International Enhancement
S.P. Agrawal Rs. 225

CICIL-37 Modern History of Punjab : A Look Back into Ancient Peaceful Punjab Focusing Confrontation and Failures Leading to Present Punjab Problem, and a Peep Ahead : Relevant Select Documents
J.C. Aggarwal and **S.P. Agrawal** Rs. 400

CICIL-38 India Information Sources : Economics and Business
Kyllikki Ruokonen and **Manohar Lal** Rs.

CICIL-39 Educational Planning in India : With a Slant to Educational Financing and Administration, Vol. 1, Reports of Committees and Commissions, Five Year Plans, Statistical Tables
J.C. Aggarwal and **S.P. Agrawal** Rs. 650

CICIL-40 Integrated Library System : Two Case Studies : Latin America and India
Maria Luiza Rigo Pasquarelli, Rosaly Favero Krzyanowski, Ines Maria M. Imperatriz and S.P. Agrawal Rs. 160

CICIL-41 Social Science Information in India: Effort towards Bibliographical Control
S.D. Vyas Rs. 300

CICIL-42 Higher Education in India Since Independence: U.G.C. and Its Approach
O.P. Gupta Rs. 360

CICIL-43 Government and Politics in India: A Bibliographical Study of Contemporary Scenario Chronicling Rajiv Gandhi Era
S.P. Agrawal and **Rajeev Kumar Sharma** Rs. 500

CICIL-44 Second Handbook of Psychological and Social Instruments
D.M. Pestonjee Rs. 300

CICIL-45 Development/Digression Diary of India; 3D Companion Volume to Information India: 1990-91
S.P. Agrawal Rs. 200

CICIL-46 Development/Digression Diary of India; 3D Companion Volume to Information India: 1991-92
S.P. Agrawal Rs. 300